H Your EALTHY PET

Baby Shampoo
3/6 wks - puppy deworm
tomato sauce - skunk odors
6 Wks - weaning
1-3 wks - cant see/hear
4 wks up - walk, teeth
2 & hrs after birth will not eat
2x increase of foods

Your HEALTHY PET

A Practical Guide to Choosing
and Raising Happier,
Healthier Dogs and Cats

Amy Marder, V.M.D.

Pet Columnist, **PREVENTION** Magazine

 Rodale Press, Emmaus, Pennsylvania

NOTICE

This book is intended as a reference volume only, not as a treatment manual. If you suspect that your pet has a medical problem, please seek competent veterinary advice. The information here is designed to help you make informed decisions about your pet's health. It is not intended as a substitute for any treatment prescribed by a veterinarian who knows your pet's need firsthand.

Cover and Book Designer: Lisa Nawaz
Cover Photographer: Cheryl Clegg

"Recipes for Pet-Slimming Diets" on page 88 was adapted from *Small Animal Clinical Nutrition III,* by Lon D. Lewis, Mark L. Morris, Jr., and Michael S. Hand. Topeka, Kansas: Mark Morris Associates, 1987.

Library of Congress Cataloging-in-Publication Data

Marder, Amy.
   Your healthy pet : a practical guide to choosing and raising happier, healthier dogs and cats / Amy Marder.
      p.   cm.
   Includes index.
   ISBN 0-87596-185-1   paperback
   1. Dogs.   2. Cats.   3. Dogs—Health.   4. Cats—Health.   I. Title.
SF427.M34 1994
636.7'089—dc20   93–27730
                                                                CIP

Distributed in the book trade by St. Martin's Press

2   4   6   8   10   9   7   5   3   1 paperback

─── OUR MISSION ───

We publish books that empower people's lives.

─── RODALE ❦ BOOKS ───

Contents

Part I: Adopting a Pet

Part II: Keeping Your Pet Healthy

Part III: Feeding Your Pet Right

Part IV: Whole-Body Fitness for Pets

Part V: The Well-Behaved Pet

Part VI: Happier Pet Ownership

Part VII: Selecting the Breed That's Right for You

PART • ONE

Adopting a Pet

Picking the Right Pet

——

The sight of an adorable kitten or puppy in a pet store or city shelter is often enough to make us pull out our checkbooks and sign the adoption papers. But too often that hasty decision leads to disappointment and regret. Many pets end up being abandoned because they don't fit in with their owners' lifestyles. When that happens, both the pets and the people suffer.

To ensure a loving and lasting relationship with your pet, choose wisely. Use your head, not your heart; consider the kind of pet best suited to the way you live, your home environment and your family needs. And while you're thinking about choices, don't limit your options to cats and dogs. For some people, birds, fish, reptiles, rabbits, guinea pigs and hamsters make more sense as pets because these animals usually are easier to care for.

PETS FOR BUSY PEOPLE

If you work full-time outside the home, think about adopting one of the easier-to-care-for animals. Cats are another good option. They are less social than dogs, so they fare better when left alone, provided they have food, water and a litter box. That means nine-to-fivers can go off to work with no worries about how their pets are doing.

Busy people, whether working at home or away, should think about which breeds require more care and attention than others. Certain types of dogs, such as retrievers and setters, for example, are highly active and need lots of exercise. Long-haired cats need to be groomed more often than the shorter-coated breeds. Samoyeds and keeshonds have thick, beautiful coats, but it takes a lot of work to keep them that way.

Short-haired dogs and cats are the best bets. Their only drawback is that some breeds shed. However, dogs such as poodles and bichons frises, which hardly shed at all, need to be clipped regularly.

PETS FOR APARTMENT DWELLERS

First and most important, check with your landlord before committing to pet ownership.

Once you have your landlord's approval, consider space constraints. Obviously, small animals are perfectly content in small spaces. Larger pets should be chosen with more attention to their intended surroundings. The idea that large dogs need large yards, however, is really an oversimplification. The important concern is activity level.

Active dogs, such as retrievers, do not do well locked in small apartments without a place to run. Great Danes, on the other hand, do not require a great deal of exercise and are fine as long as they get out to stretch their legs regularly. Cats are considered the perfect apartment pets. But cooped up in small spaces, they often scratch the furniture. If you adopt a cat, provide it with an appropriate scratching post. Real logs and sisal-covered posts will distract Kitty from your precious furniture.

PETS FOR PEOPLE WITH CHILDREN

If you have children and a good-size fenced-in yard, a playful and active dog may fit well into your lifestyle.

Collies, golden retrievers and shelties are active dogs that tend to be good with children. Just remember: Personality traits are trends only; individuals vary within each breed. If the animal is a previously owned adult, be sure it has been exposed to children in the past.

Consider, too, the age of your child. Very young children don't know how to be gentle. A child isn't ready for a pet until he can tell the difference between right and wrong and has grip control. I would not introduce a pet into a home where there are children under three or four years of age.

Finally, never agree to buy a pet for a child with the stipulation that the child must care for it. Almost invariably, pet-care responsibilities fall to the parent, usually the mother. You must be willing to accept them.

TIPS FOR PICKING A GOOD PET

Beyond those specific lifestyle considerations, here are some basic tips for choosing a good pet.

• Learn as much as you can about the characteristics of different animals and breeds, so you can narrow your choices to those that best fit your lifestyle and home environment. A good place to learn about breeds is at dog and cat shows. Try to speak not only to the breeders but also to the handlers. Many handlers are knowledgeable about several breeds and can give you valuable information. (You might also want to refer to chapter 41, which compares the most popular breeds of dogs and cats.)

• If you decide to adopt a puppy or kitten, choose one between six and eight weeks of age. This is the critical period during which socialization to both people and other animals is most important.

• Choose your pet from a healthy litter. Try to meet your prospective pet's parents. Ideally, these mature animals should be healthy and well adjusted (not fearful or aggressive). Also, look for an animal raised in a household where good nutrition, kindness and attention are the rule.

• When choosing, try to steer away from both the most aggressive animal in the litter and the shyest. Stick to animals with middle-of-the-road temperament. Although many people are especially attracted to the runt of the litter, it's a fact that these animals stand a greater chance of having future behavior problems than do their siblings.

• If you prefer to skip the chewing/housetraining stage and adopt an adult dog (or cat), try to obtain a good history on the animal. This is important because some adult animals bring quite a bit of "baggage" with them. Many are up for adoption because of serious behavior problems that may not surface until you have them in your home. Observe your potential pet carefully. Make sure it is friendly to all members of your household. If it is a dog, ask if it's been housetrained; if it is a cat, ask if it uses the litter box.

• Consider the gender of the animal. In general, male dogs and cats have more behavior problems than females. Males of most species tend to be more aggressive than females. They also engage in urine marking (spraying in cats) and mount more than females. Although often misunderstood, protectiveness and territoriality are not gender-related traits. Both males and females can make excellent watchdogs.

• Try to arrange the adoption on a trial basis. Work with your new pet for a couple of weeks before making a final decision. Shelters, pet stores and breeders who are really sincere about placing pets in good homes will be happy to take back the animal if things do not work out.

HOW TO SHOP THE SHELTERS
FOR YOUR NEXT PET

If you decide on a particular breed of dog, among the best places to start are the breed rescue leagues. You can locate them through the American Kennel Club (51 Madison Avenue, New York, New York 10010). Rescue leagues tend to have the most information on the history of the dogs and are good at steering you toward the best dog for your family.

The city and humane society shelters are excellent sources of mixed breeds and cats. Many city and county shelters are well organized and supervised by animal control officers. But some are poorly run. To get more information about shelters in your area, call your local animal control officer.

Then check out the shelter yourself. The animals should look well cared for. (If you're not permitted to see the area where the animals are kept, assume the worst.) And the shelter staff should ask you questions to determine whether you are prepared to care for the pet you select.

Many shelters run by humane societies have adoption counselors who may have information on the animal you want to adopt. They also observe the animal's behavior, so they can give you sound advice on whether you'd make a good match. Many also have veterinarians on staff who can give the animal a health exam.

But call first to see which shelters have the type of animal you want.

PICKING OUT YOUR PET

Explore the animal's history, health and behavior before you take it home. A good shelter or adoption agency may want to interview you to make sure you can provide a successful home for the pet. Here's how to investigate a potential pet's personality.

Run a background check. Where did the pet come from? Were there children or other pets in the family? Did the pet spend time outdoors? How many homes did it have before? Why was the animal given up? The staff may not be able to answer all these questions, but you should have this information before you make your choice.

Gather staff observations. How long has the dog or cat been up for adoption? Has it ever been adopted and returned? How does the animal behave with the staff, other animals and visitors? Has it displayed any health problems or unusual habits, such as chewing or scratching on objects?

Make some observations of your own. Does your potential pet look healthy, bright and alert? When you approach the cage or run, does the animal act friendly? Or does it back up and growl, bark or hiss? Although you may feel sorry for animals that appear to be afraid, a friendly animal is a better bet. If you are adopting a cat, look at the litter box and make sure it is being used.

Take the pet for a test drive. Ask an attendant to take the animal out of its cage, so it can roam or run. How does the animal behave when you reach for it or pick it up? Does it try to bite or scratch? If there are other animals around that it may be reacting to, try to take it into a private room. If the aggression subsides, that's a good sign. If not, look at another animal.

Take out some yarn (for a cat) or a dog toy and see if you can get the animal to play. An animal that plays in this situation is relaxed and confident. If it doesn't play, though, it may just need some time to warm up to you.

If it's a dog, is it easy to put on a leash? How does it act when you take it past other dogs and people? Barking at other dogs can probably be controlled, but aggression directed at you or other people is not good.

Talk it over with the members of your household. Take the time to discuss the adoption fully with your family. Make sure each member gets to meet your potential pet before you finalize the deal. That can head off any problems individuals may have.

As you can see, there are many factors to consider in choosing a suitable pet. Taking a little extra care and giving the necessary time to the process will pay off handsomely. You will have the joy of a compatible companion and avoid the disappointment or annoyance of dealing with a pet that doesn't suit your lifestyle or personality or live up to your expectations.

· 2 ·

Choosing a Veterinarian

___

Most caring pet owners seek to form long-lasting, reliable and trusting relationships with good veterinarians. But like the many types of doctors who take care of humans, there's a wide variety of doctors who treat animals. Not only do veterinarians vary in their specialties (such as exotic animals, cardiology, surgery), they also vary in education, personality, treatment philosophy, equipment, services available, support staff and fees. So how do you go about choosing the best veterinarian for both your pet's needs and your own?

Make a list of all the things that are important to you in a veterinarian. It may include the location of the office itself, the range of services you want and specific characteristics you'd like to see in the person who will be your vet.

THE OFFICE

You probably want a veterinarian who's close by and easily accessible. You're much more likely to take your pet to see a doctor who's only a few minutes away than to one whose office is on the other side of town. Easy accessibility makes those very important preventive well-pet visits easier, and it may be lifesaving in an emergency.

Think twice, however, before you settle for the very closest veterinarian (as many people do). Check around. There may be another one just a few blocks away that suits you better.

Is the exterior of the office well kept? How about the inside? An untidy office may signal a careless and not-so-thorough doctor.

Are there separate waiting rooms for dogs and cats? Going to the veterinarian is a bad enough experience for most cats, and a waiting room full of dogs just makes matters worse.

THE SERVICE

These days most veterinarians keep hours to accommodate pet owners who work outside the home. Evening and Saturday hours are commonplace. Many also offer a drop-off service: You leave your pet in the morning with a list of everything it needs and pick it up after work. Some veterinarians also make house calls, in case it is difficult for you to get to the office.

How many veterinarians are on staff? Are you usually able to get the one you like, or do you frequently have to settle for the covering doctor? Is the support staff trained and licensed? Although some states don't require veterinary nurses to be licensed, you know that those who are licensed have achieved a certain level of competence. Their training and knowledge enable them to take better care of your pet. Above all, is the staff friendly and caring?

Although some veterinarians still see all their own emergency calls, most share emergency coverage with other veterinarians or send their patients to an emergency clinic or referral hospital (such as at a veterinary school). But no matter what the arrangement, it's essential that, in an emergency, your pet be seen promptly by a competent veterinarian.

In my opinion, emergency clinics or hospitals are able to provide the best and quickest service. Most clinics have veterinarians on the premises at all times, plus a well-trained support staff and all the equipment needed to handle most emergencies. Generally, you are allowed to transport your pet to your regular veterinarian the next day for any necessary ongoing care. It's very important that the veterinarian you choose have a dependable emergency service you feel comfortable with.

Some veterinarians run clinics that handle mostly outpatients (vaccines, simple medical problems) and simple surgeries (such as neuterings and dental work). These doctors usually refer serious medical problems and difficult surgeries to larger hospitals equipped to handle such cases and to provide the special care that sicker animals need. When choosing your veterinarian, make sure you ask if animals can be hospitalized on the premises if they need to be. And if so, is there someone in the building at all times? (Many times no one is there.)

Ask about equipment. This varies quite a bit. For instance, not all veterinarians have x-ray machines. What about a sterile surgery? Does the vet use gas anesthesia (which is probably the safest type)? Does the office do its own lab work, or is it sent to an outside lab? (It usually takes longer to get results if it is sent out.)

Just as there's a wide variety of services a veterinarian may offer, there's also a wide range of fees charged for similar services. If you shop around, however, make sure you consider the quality of the service as well as how much it costs. An inexpensive spaying procedure done in an unsterile surgery suite with unsterile instruments may result in infection and a lengthy recovery time.

Some veterinarians, especially in the western part of the country, are now able to offer pet health insurance. Make sure you ask about its availability in your area. It may make paying for your pet's medical care much easier.

THE DOCTOR

Now let's get to the most important issue: the person who will be your vet. It's essential that both you and your pet get along well with that person.

A few things to think about before actually meeting your potential doctor: Do you prefer a man or a woman? (Does your pet seem to do better with men or women?) Do you want an older veterinarian with several years of experience, or do you prefer a recent graduate, who may be more familiar with all the newest techniques?

After you meet the doctor, observe how he handles your pet. How does Fido or Fluffy react? Does the veterinarian take the time to answer your questions completely? It may be important to you to find out how the doctor feels about unconventional therapies, such as acupuncture. How about views on animal rights and the proper time for euthanasia? And if you introduce your pet as a family member, what response do you get?

What are the veterinarian's special interests? Does he have any expertise in treating pets such as yours? (Cats are not little dogs, and birds are in a class of their own.) And very important, how does he feel about referring cases to a specialist should the need arise?

STARTING THE SEARCH

Many people look for a veterinarian in the Yellow Pages. Although the phone company might disagree, this is probably the least desirable way to find a good veterinarian. Such an ad is often merely a reflection of how much money the veterinarian is willing to pay for advertising. Another common and equally risky method is to choose solely by the vet's proximity to your house.

Better ways to begin your search are by word of mouth and through veterinary associations. Talk to pet owners you know, local dog club or cat club members and obedience instructors. Steer clear of advice from pet stores, as they are often connected financially with a particular veterinarian.

The American Veterinary Medical Association (1931 Meecham Road, Suite 100, Schaumburg, IL 60173) can refer you to state veterinary associations, which will be able to help you. The American Animal Hospital Association, or AAHA (P.O. Box 150899, Denver, CO 80215), is a group that accredits animal hospitals meeting certain high standards. In general, any hospital that displays a certificate saying it is an AAHA member is well equipped. The AAHA will be able to tell you of any member hospitals in your area.

Another recently formed group that can help you choose a veterinarian is the American Board of Veterinary Practitioners, 530 Church Street, Suite 300, Nashville, TN 37219. The purpose of this board is to certify veterinary practitioners, by examination, who "fulfill the specific, functional role of delivering superior, modern, comprehensive veterinary services." A board-certified practitioner will be knowledgeable and up to date.

After you pick a person on paper, the next step is to schedule an appointment. Make it for a time that allows for an interview and a tour of the hospital and facilities.

If you choose a veterinarian carefully, your effort will be rewarded each time you take your pet to the clinic. That happy tail wag or contented purr lets you know that you made a good choice.

• 3 •

The Great Value of Vaccinations

___

One of the most important responsibilities of adopting and owning a dog or cat is making sure that the animal has the vaccinations it needs. So whatever the age of your prospective pet, be sure to look into its immunization history.

All dogs and cats need to be immunized. Even pets that rarely or never go outside can be exposed to disease-causing viruses carried in the air or on our clothes or shoes. Many people erroneously believe that older pets no longer need to be vaccinated. But these, along with very young animals, are the most susceptible to disease.

Vaccines are modified or killed forms of disease organisms that stimulate the immune system when introduced into an animal's body. Vaccines are usually injected but are occasionally given through the nose or by mouth. Your pet produces antibodies when it receives a vaccine. Then, if your pet is exposed to the actual disease later, these antibodies kill the infectious agent. For optimal protection, antibody levels must be high. So it's very important for your pet to receive periodic booster vaccinations.

The best time to begin your pet's first series of vaccinations is when the animal is 6 to 8 weeks old. Before then, your puppy or kitten is protected from disease by the antibodies it receives from its mother (if the mother was vaccinated). Since it is difficult to predict exactly when these antibodies will be depleted, the initial immunization regimen is divided into a number of doses, usually given 3 to 4 weeks apart. The last vaccine in the puppy or kitten series is usually given at 14 to 16 weeks of age.

The benefits of any vaccine far outweigh the risks. Vaccination should be postponed, however, if the animal is very ill, pregnant or being treated with drugs known to suppress the immune system.

YOUR DOG'S VACCINATIONS

The vaccines for canine distemper, infectious canine hepatitis, leptospirosis, parainfluenza and parvovirus are usually combined into a single vaccination called DHLP-P. This combination, used to vaccinate puppies and as an annual booster for adults, is very effective. The vaccine for rabies is given separately to puppies according to the schedule described below.

A vaccine for Lyme disease, also available now, is given separately as well. Other preventives, generally used at the veterinarian's discretion, protect dogs against *Bordetella bronchiseptica* and heartworms. These are also given separately.

Distemper. This widespread, highly contagious viral disease is usually fatal. Symptoms include fever, coughing, sneezing, runny eyes and nose, vomiting, diarrhea and loss of appetite. Ultimately, the virus attacks the dog's nervous system and may cause seizures.

Infectious canine hepatitis (ICH). Another viral disease, ICH primarily affects the liver. Signs include fever, lethargy, abdominal pain and a blue haze over the dog's eyes.

Leptospirosis. This bacterial disease is transmitted through the urine of an affected animal. It may infect the dog's kidneys and cause weakness, vomiting, high fever and loss of appetite.

Parainfluenza. This virus is part of the group of viruses and bacteria responsible for infectious tracheobronchitis, or kennel cough. Characterized by a dry, hacking cough often accompanied by fever and loss of appetite, kennel cough is a highly contagious airborne disease. Vaccines may help prevent it.

Parvovirus. A relatively new disease, parvovirus usually causes vomiting, bloody diarrhea and loss of appetite in dogs of all ages. If untreated, affected dogs often become dehydrated and die. The virus may also affect the hearts of very young puppies and cause death within 12 hours.

Rabies. This deadly viral disease affects the nervous systems of all mammals, including man. It is transmitted mainly by a bite from a diseased animal. Strains of the rabies virus have become established in wild animals—mainly skunks, foxes, raccoons and bats. Dogs and cats, however, are the main carriers of rabies to humans.

Diseased animals may become highly excitable (furious rabies) or paralyzed (dumb rabies). Animals in the furious stage become very sensitive to noise and often bite indiscriminately. Because of the public health dangers, vaccination against rabies is mandatory almost everywhere. Your puppy's first vaccinations should be given at three to six months of age and again one year later. After that, boosters should be given annually or every three years, depending on which vaccine your veterinarian uses and your state or local laws.

Lyme disease. Carried by ticks, this disease has reached epidemic proportions in some areas of the United States. Dogs with Lyme disease have a fever, diminished appetite and arthritis. If you live in an area where Lyme disease is prevalent and your dog has access to tick-infested areas, ask your veterinarian about the possibility of protecting your dog with the new vaccine. (Since we're not sure how long the protection lasts, it's important to continue to use tick insecticides and repellent as well as to check for ticks daily.)

Bordetella bronchiseptica. Common to the respiratory tracts of many animals, these bacteria are a primary cause of kennel cough. Some veterinarians recommend vaccinations for all dogs; others reserve the vaccination for those dogs at high risk (show dogs and dogs being kenneled, for example).

Heartworms. A preventable disease of dogs (much less common in cats), heartworm disease is caused by the worm *Dirofilaria immitis,* which is carried by mosquitoes. Infected dogs tend to tire easily with exercise, cough and lose weight. If the disease is prevalent where you live, your veterinarian will prescribe pills to use during the mosquito season as a preventive. The pills may have bad side effects in dogs that are infected with the worms, so your veterinarian will also recommend a yearly blood test to be sure that your dog is clear.

YOUR CAT'S VACCINATIONS

The feline vaccines for viral rhinotracheitis, calicivirus and panleukopenia are usually administered in a combination often called FVRCP. This vaccine, used to immunize kittens, can also serve as a yearly booster for adults. The vaccinations for feline pneumonitis, rabies and feline leukemia are given separately but also need to be included in the kitten series and then repeated regularly as boosters.

Feline viral rhinotracheitis (FVR). A common, highly contagious viral respiratory disease, FVR may cause fever, depression, sneezing, coughing and loss of appetite. Vaccines are available in both injectable and intranasal forms.

Feline calicivirus. Another viral respiratory disease, this one not only may cause the same symptoms as FVR but also may produce painful ulcers in the cat's eyes, mouth and throat. Vaccination may be given either by injection or by the intranasal route.

Feline panleukopenia. This highly contagious and often fatal viral disease of cats is particularly deadly to kittens. Also called feline distemper or feline infectious enteritis, it usually causes fever, vomiting and diarrhea, resulting in severe dehydration. The virus also attacks the cat's bone marrow and depletes immunity. Vaccination results in rapid and effective protection.

Feline pneumonitis. Some veterinarians also vaccinate for pneumonitis, another respiratory disease of cats. The vaccine may be given separately or in combination with the other feline vaccines.

Rabies. There are approximately 200 feline rabies cases per year in the United States. Since these cases pose a significant risk to humans, the American Veterinary Medical Association recommends that all cats be routinely vaccinated for rabies.

Kittens should receive their first vaccines at three to four months of age and then one year later. After that, boosters should be given annually or every three years, depending on which vaccine your veterinarian uses and on your state or local laws.

Feline leukemia. This virus, one of the leading causes of death in our cat population, can result in a variety of problems besides leukemia. Affected cats may develop anemia, solid tumors and impaired immunity. Cats with lowered immunity often develop other chronic ailments, such as respiratory diseases and mouth infections.

A vaccine providing protection from this deadly disease was introduced in 1985. Most veterinarians recommend vaccinating only those cats that are likely to be exposed to the virus. Specifically, cats that go outdoors and cats that live in households with many other cats are most at risk.

Before beginning the vaccination program, your veterinarian will take a blood test to make sure that your cat is not a carrier. The first dose of vaccine can be given any time after your cat is nine weeks of age, the second two to three weeks later and the third two to four months after the second. After the third, yearly boosters are required to maintain adequate immunity.

Nothing you can do to preserve the health of your pet is more effective than vaccination.

Why It Makes Good Sense to Get Your Pet Fixed

To neuter or not to neuter? It's a difficult decision for many new pet owners. Well-meaning friends and relatives often complicate matters by providing misinformation on the subject. However, the facts show that neutering is a caring step you can take to ensure a happier, healthier life for your dog or cat while preventing overpopulation. Here's why.

NEUTERING IS BETTER FOR THE ANIMALS

Of course, your primary concern is for the pet itself. The advantages for your dog or cat are hard to ignore.

It can prevent disease. Neutering is the permanent sterilization of an animal. In females, the procedure is often called spaying and is the equivalent of a total hysterectomy (removal of the ovaries and uterus). It eliminates the chances of uterine infection or cancer. It also greatly reduces the risk of breast cancer.

In males, neutering is usually referred to as castration or altering and involves the removal of the testicles. It eradicates the possibility of testicular cancer or infection. In dogs, it also prevents the occurrence of serious and painful prostate problems, common in older males.

It saves lives. In one recent year, over 17 million unwanted dogs and cats were dropped off at our nation's animal shelters. And unfortunately, only about one in six of these orphaned pets found a new home. The rest, some 13.5 million (most young and healthy), had to be put to sleep. The only way to halt the tragedy of pet overpopulation is to stop allowing our pets to breed. And the best way to prevent breeding is to neuter.

It helps the homeless. If you don't let your pet breed, you increase the chances for homeless pets. There aren't enough homes for all the pets in our shelters right now, let alone those not yet born.

IT'S EASIER ON THE OWNER

There's nothing wrong with enjoying the benefits that come to you, the owner, as a result of neutering your pet. Consider, then, a bonus for your good deed.

It eliminates unwanted behaviors. In males, neutering eliminates most of the sex hormone testosterone, which is largely responsible for an animal's "male behaviors." In dogs, these include roaming, mounting, urine marking (leg lifting) and fighting with other male dogs. Neutered males have a decreased desire to roam and search for females and therefore are less likely to be injured or killed in auto accidents or to be lost. After a male is neutered, he not only is less likely to pick fights with other male dogs but also is less likely to be attacked.

In cats, too, altering decreases or eliminates roaming, mounting, spraying (urine marking) and fighting. Contrary to common belief, though, castration does not make an animal calmer. Only maturity, exercise and good training can calm a very active young dog or cat.

Spaying has much less effect on female behavior. However, it does prevent the irritability and occasional aggressiveness displayed by some female dogs and cats during heat (their fertile period) and by some female dogs during the period of false pregnancy that follows heat.

It makes animals easier to care for. Unneutered males tend to have only one thing on their minds if there are females around outside. Trying to keep them indoors may be difficult, as they become restless, overactive and often destructive.

Because spaying eliminates a female dog's heat cycle, you won't have to contend with furniture and carpet stains from the bloody discharge that occurs during that time. And spaying puts an end to the persistent serenading of your female pet by neighborhood males.

IT'S SAFE AND BENEFICIAL

The long-range results of neutering are equally attractive for your pet and for you.

It helps pets live longer, happier lives. Because they stay healthier, neutered pets live longer. They're also happier, as they have an easier time adapting to life with humans.

It has no ill effects. Neutering won't turn your watchdog into a wimp. The procedure has no effect on a dog's basic instinct to defend its territory.

Neutered pets do not become psychologically depressed either, as one common myth asserts. In fact, most neutered pets are more loving and responsive to their owners, since they are no longer preoccupied with mating.

And contrary to popular belief, neutered animals do not become fat and lazy. Neutered pets do require fewer calories, but when placed on a proper diet, they maintain their trim figures, provided they continue to exercise regularly.

The benefits outweigh the risk. Although neutering is a surgical procedure that does require general anesthesia, new developments in the field of veterinary medicine have greatly reduced the risk involved. Certainly, the very minor risk is far outweighed by the benefits.

It's fiscally sound. Most humane societies make neutering affordable by offering low-cost programs. If you think you can make money selling your pet's offspring, be forewarned. As most breeders can tell you, if you provide the puppies or kittens with proper care, you'll be lucky if you break even. And if you decide to keep one, the price of raising it quickly surpasses the savings from not neutering the mother.

It's never too late. Although the best time to neuter your pet is between six and eight months of age, the surgery has the same effect on behavior and reproduction no matter when it's done.

PART TWO

Keeping Your Pet Healthy

· 5 ·

A Pet Owner's Guide to Sorting Out Symptoms

—

Helping your pet live a longer and healthier life requires not only a balanced diet and plenty of love and play but also the early detection of disease. Many pets are saved by observant owners who take them to the veterinarian at the first signs of illness. Sometimes, however, it's difficult to tell if the signs your pet is showing are something to worry about. So here's a rundown of some of the most common symptoms your pet may show, what they might mean and what action you should take.

WATCH THEIR WEIGHT

Weight loss. If your pet is eating its usual amount of food but is losing weight, that's a signal that something may be wrong. Your pet may have a gastrointestinal or metabolic disorder such as hyperthyroidism, a condition common in older cats. Since it can be difficult to judge an animal's weight visually, it's a good idea to weigh your pet once a month and keep records. Generally, a loss of one pound in a cat or small dog or two to three pounds in a larger dog is reason to consult your veterinarian.

Weight gain. If your animal is eating normally but gaining weight, that's a reason to call the doctor, too. A sudden increase in weight may indicate fluid accumulation and may mean that your pet has underlying kidney, heart or liver disease. Or if your pet is gaining weight and seems especially lazy, it may have hypothyroidism, a common canine disease. Of course, eating high-calorie junk foods can cause weight gain, too. If that's the case, try to feed your pet only high-quality pet foods.

Loss of appetite. Appetite loss, or anorexia, occurs in a wide variety of diseases. If your pet doesn't eat for more than a day, call your veterinarian.

Excessive appetite. Overeating, or polyphagia, can occur in diseases that prevent the absorption and utilization of nutrients. Diabetes and gastrointestinal problems are examples. An increase in appetite also occurs when an animal needs more food, such as in pregnancy or when intestinal parasites compete for the food it has eaten. In cats, hyperthyroidism frequently causes a voracious appetite. If your pet overeats for more than one day, consult your veterinarian.

WHY IS MY PET ACTING THIS WAY?

Fever. Most owners can tell when their pet has a fever because the animal is inactive and has a poor appetite and a "dull" look in its eyes. (Contrary to popular belief, your dog's nose is not a good indicator. Feverish dogs may have either a cool, wet nose or a warm, dry one.) To take your pet's temperature, put a dab of petroleum jelly or lubricating jelly on the tip of a rectal or digital thermometer and insert it about one inch into the rectum. Read it after one minute. If the reading is over 103°F, your pet has a fever. See your veterinarian, so she can determine the cause.

Pain. This can be one of the earliest signs of disease. Animals in pain may become listless, shiver, groan and whimper, move constantly or be reluctant to move. Neck or back pain often causes animals to avoid moving, and if forced to move, they often cry out. Abdominal pain may cause them to assume a "praying" position, with hind legs upright while lying down with the front legs. Any type of severe pain should be checked immediately by your veterinarian. Mild pain that doesn't go away in a day should also be checked.

INTESTINAL UPSETS MUST BE MONITORED

Vomiting. Viral diseases, intestinal obstructions (hair balls, tumors, strings) and kidney and liver disease can all cause vomiting. So can something as simple as an abrupt change in diet. As vomiting can be very debilitating and can quickly lead to fluid imbalances, call your veterinarian if your pet vomits more than twice in one day.

Diarrhea. The most common cause of diarrhea is a sudden change in diet—as when table scraps are fed or when your pet gets into the garbage. This type of diarrhea is usually short-lived and is not serious (although preventing it is a good idea). Intestinal parasites also cause diarrhea, especially in puppies and kittens. Have your young animal's stool checked regularly. Other common causes of diarrhea are infectious diseases (such as parvovirus), allergies, digestive disorders, kidney and liver disease and intestinal cancer. Whenever diarrhea lasts for more than one day, take your pet and a sample of its stool to your veterinarian.

Constipation. It is very important to monitor your pet's feces on a daily basis, so you know what is normal for your animal. Note any changes in the color or consistency of the stool. Any change in bowel habits, fecal color or consistency that persists for more than one day is of concern.

Blood in stool. This is always a cause for concern. Notify your veterinarian as soon as possible.

PERSISTENT COUGHING AND SNEEZING

Coughing. In dogs, coughing is often caused by infectious tracheobronchitis, or kennel cough. Cats often cough in an attempt to expel a hair ball. Other more serious causes of coughing are pneumonia, abnormalities of the trachea (windpipe), heart disease, allergies, parasites and cancer. Coughing should always be checked by your veterinarian as soon as possible, especially when it's accompanied by labored breathing.

Sneezing. Infectious upper respiratory diseases are the most common cause of sneezing. Occasionally, though, plant materials, such as foxtails, are inadvertently sniffed into a nostril and cause a severe bout of sneezing. A couple of sneezes a day may be normal, but persistent sneezing or sneezing accompanied by nasal discharge is not. See your vet.

WHAT DRINKING AND URINATION TELL YOU

Excessive fluid intake. On an average day, your dog or cat should not be drinking more than one ounce of fluid per pound of body weight. (It should be urinating about half that amount.) If you are feeding your pet dry food now or switch from canned to dry or semimoist food, the animal will drink more water to fulfill its needs. That's normal. Any other change in drinking habits that lasts for more than two to three days should be checked by your veterinarian.

Excessive urination. If your pet is having urine accidents in the house, is urinating more than usual, is having difficulty urinating, has bloody urine or doesn't seem to be able to control urination, take the animal and a urine sample to your veterinarian as soon as possible. Your pet may have a serious kidney or bladder problem.

CHECK THE EYES AND EARS

Eye symptoms. A small amount of clear discharge from your pet's eyes is probably normal (if it is not causing any discomfort). Any change in the amount or type of discharge warrants a call to your veterinarian. Cloudiness, pain (often indicated by squinting), itchiness, redness or discomfort of any kind should be checked as soon as possible. Loss of vision may be the problem if you notice your pet bumping into recently moved or new furniture.

Ear symptoms. Pets with ear infections often paw at their ears, shake their heads and rub their ears on the ground. There may be a discharge or a noxious odor. Ear infections are very uncomfortable, so take your pet to the doctor as soon as possible.

DANGER SIGNS IN THE MOUTH

Gum symptoms. Your pet's gums should be pink. Pale gums may mean that your animal is anemic; bluish gums are a sign of serious respiratory or heart problems; and a yellowish hue may indicate liver problems. Black spots on your pet's gums may just be normal pigment spots, but it's best to have any change in gum color checked by your veterinarian.

Yellow teeth. Your pet's teeth should be white and smooth. If they are yellow or brown, they are probably covered with dental tartar and need to be cleaned.

Bad breath. The most common cause of bad breath is dental disease, but it can also be caused by kidney disease and digestive problems. Have your veterinarian check it out.

KEEP AN EYE ON SKIN AND BONES

Skin symptoms. Excessive scaling or dandruff or a bad odor may be a sign of skin disease. Itching and scratching may be caused by fleas, allergies, infections, dryness or external irritants. Hair loss may reflect underlying medical problems, such as hypothyroidism in dogs or Cushing's disease. Run your hand through your pet's coat daily. If you notice any lumps, have them checked by your veterinarian, especially if they seem to be growing.

Stiffness, lameness or swelling. If your pet is stiff when it wakes up, limps or has a swollen leg, a joint or bone problem may exist. Have your animal checked if its symptoms persist for more than one day.

Behavior changes. Any marked change in behavior that persists for more than two days is probably significant. Have your pet checked out medically.

Seizures or convulsions. These are always a cause for concern. See your veterinarian immediately.

Remember, you are the best observer of your pet. If you notice anything that concerns you, see your veterinarian. Just as in people, there is a better chance of curing a disease if it is caught and treated early.

• 6 •

Protect Your Pet with Health Exams at Home

M ost pets are seen by their veterinarians only once a year for vaccinations and a physical exam. Sometimes a veterinarian will find something abnormal during the exam, such as a lump, and will need to know some history in order to evaluate its seriousness—how long has the lump been there? Has it changed in shape or color? If your pet has lost weight, the doctor will ask if the loss is recent or part of a steady weight loss over the year.

DON'T DEPEND ON ANNUAL CHECKUPS TO CATCH EVERYTHING

An annual checkup by your veterinarian is an absolute necessity, but that may not be frequent enough to detect the early signs of a disease. You can perform a valuable service for your pet by conducting health examinations—at least weekly—at home. In addition to protecting your pet, the touching and massaging you do during the exam relaxes your pet and strengthens the bond between the two of you.

Plan to carry out your home health exam at the same time you groom your pet. Begin performing this ritual as soon as the animal is weaned and comes under your care. Try to make the sessions enjoyable, like a game; that way, your pet will learn to like being handled, instead of being fearful. Enhance the experience with food treats, toys and gobs of praise for your pet. Schedule these sessions for the animal's calmest times. Never attempt to examine or groom an immature pet during an active play period—in dogs, fondly known as the "puppy nutties."

START WITH A QUICK ONCE-OVER

Begin the examination by making a mental note of your pet's overall attitude. Is the animal behaving normally or acting listless and a little under the weather? Then begin to massage your pet's entire body. It doesn't really matter whether the animal is standing, sitting or lying down; just try to include all the body parts. Begin with the head and neck area, including the ears and muzzle. Then gently move across the back, sides and underside to the tail.

As your hands travel along, feel your pet's ribs. Is there an obvious fat layer over them? If your answer to this question is yes, you should see your veterinarian about starting a weight reduction program for your pet. (If your pet's ribs are much more prominent than usual, it may be due to abnormal weight loss resulting from an ailment, and you should consult your veterinarian about that as well.) Of course, weighing your animal on a scale once a month is an even better way to keep track of its weight.

Check out the legs and, finally, the feet and toes. If your pet resists your massage, be patient and gentle but persistent. Most pets will start to relax and give in. (But some animals might growl or hiss; if this happens, do not continue. Contact your veterinarian for assistance.)

Use the massage as a way to become familiar with your pet's body. Pay attention to normal lumps and bumps, so

20 Warning Signs
That Demand Professional Attention

If you see any of the following signs as you observe or examine your pet at home, consult your veterinarian at the first opportunity. Each of these symptoms is explained in more detail in chapter 5.

1. Poor appetite for more than one day

2. Increased appetite for more than one day, especially when accompanied by weight loss

3. Weight loss when food is not restricted, especially if food intake is normal or more than normal

4. Weight gain without increased food intake, especially if food intake is less than normal

5. Fever

6. Pain

7. Vomiting more than twice (immediately if the vomit is bloody or dark)

8. Diarrhea for more than one day (immediately if the fecal matter contains blood)

9. Any change in bowel habits, fecal color or consistency for more than one day

they won't alarm you later. Know what the ribs and bones feel like. Count your pet's nipples, so you won't be like a client of mine who brought her pet to me for nipple removal—because she didn't think her male dog should have lumps on his chest!

Assess your pet's hydration status by gently pulling up the skin on its neck. If the animal is not dehydrated, the skin will rapidly fall back into place. If your pet is dehydrated, however, the skin will remain elevated and appear a little stiff.

10. Coughing more than twice (immediately if the coughing is accompanied by labored breathing)
11. Sneezing more than twice (immediately if the sneezing is accompanied by nasal discharge)
12. Excessive drinking (more than one ounce per pound of body weight daily) for two to three days
13. Increased urination or sudden accidents in the house, difficult urination with straining, bloody urine
14. Eye discharge for more than one day (immediately if discharge is puslike); eye cloudiness, squinting, itchiness, redness, pain or discomfort; loss of vision
15. Ear scratching, head shaking and ear rubbing, ear discharge
16. Trouble with eating, such as mouth pain
17. Itching and scratching a part of the body for more than two days
18. Lameness for more than one day if not severe (immediately if lameness is severe and very painful)
19. Seizures or convulsions
20. Behavior changes that continue for more than two days

Now run your hands through your pet's coat from face to tail to detect any flaking, scabs or parasites. The parasites most commonly found are fleas and ticks. Fleas, hard brown insects that hop, may be carriers of tapeworms and cause severe skin allergies. They feed on blood and leave behind "flea dirt," which looks like coarse black pepper but turns red when wet. Ticks also ingest blood. They look like small hard spiders when not feeding and gray grapes with legs when engorged with a blood meal. Ticks may carry serious diseases that can be transmitted to both pets and people.

TAKE MORE TIME
TO CONCENTRATE ON SPECIFICS

Now do a detailed evaluation of all your pet's parts.

Look into your pet's eyes. Expect them to be bright and wide open. Check for redness or a mucous or puslike discharge. Scratches of the cornea cause pain and squinting; cataracts result in whiteness of the lens; and glaucoma symptoms include red eyes and often pain and discharge from inflamed eyelid tissues.

Examine the inside of the ears. An unpleasant smell, swelling, tenderness and discharge are all red flags. Early treatment can prevent pain.

Check inside your pet's mouth. Just as changes in skin color can warn of health problems in humans, pale gums in your pet can mean anemia; bluish gums, respiratory or heart ailments; yellowish gums, possible liver trouble; and red, irritated gums, dental disease. Check any color change with your veterinarian.

While examining your pet's mouth, search all surfaces for any lumps. (Make sure you know what your pet's tongue normally looks like.)

Yellow or brown teeth probably have a layer of tartar; a cleaning can whiten them.

Smell the animal's breath. If your pet's breath almost knocks you out, it's probably due to dental trouble. But kidney disease and digestive problems can also cause mouth odor in your dog or cat.

Don't ignore your pet's nose. Among healthy pets, the nose is not always wet and black. Some pets have pink or brown noses, and noses are often dry. Nasal discharge that lasts for more than one day usually means trouble, so have it checked as soon as possible.

Proceed to the chest area. You can feel your pet's heartbeat by putting your hand over the chest area right behind the front legs. If you feel it, count the number of beats per minute: The normal rate for a dog is 70 to 160 beats per minute (up to 220 in puppies); for a cat, it is 110 to 240 beats per minute. The beats usually occur in an even rhythm, but in some animals the heart rate speeds up and slows down with breathing. If you notice any irregularity, have your pet checked out immediately. You may also try to feel your pet's pulse on the inside of the rear leg, close to the body. This is especially difficult in cats, but with practice you'll probably be able to do it. There should be one pulse for each heartbeat.

And what about your pet's breathing? Is it regular? Does it seem difficult or unusually rapid? Is it noisy? Rapid, difficult, raspy or shrill breathing suggests a problem. Panting is normal when pets are hot, excited or stressed.

Examine the abdominal area. Notice its shape. A potbellied appearance can mean intestinal parasites in young animals or fluid accumulation, metabolic disease or tumors in older animals.

Look over your pet's legs. Notice any swellings or other areas that might be painful. Pets get many types of arthritis, bone tumors, fractures, growth abnormalities, sprains and torn ligaments, just as humans do. Observe your pet's gait. Know what is normal for your individual animal, so you will be able to detect the first signs of a limp. Check your pet's nails to see if they need to be trimmed. You should trim them yourself about every two weeks. Make sure you don't forget the dewclaw, or thumbnail.

Now examine the tail end. The anus is a common site for lumps in older animals, so it is an important part of your examination. You won't be able to see the anal sacs, but they are important to know about. They are situated on either side of the anus, just inside the anal opening, and normally discharge when an animal defecates or when it is stressed. Unfortunately, the sacs often become plugged up (in both dogs and cats), and the result is discomfort. Your pet will probably drag itself along your carpets on its hind end when an anal sac problem is present. Also, look to see if there is a discharge from the vulva (especially if it is pus-like in unspayed females) or the penis. If your pet still has testicles, it is also important to feel these for any lumps or changes in size.

If your pet seems ill, take its temperature. Animals that have fevers are usually depressed and have poor appetites and lifeless eyes. (Your dog's nose—warm or cool, wet or dry—is utterly unreliable as a fever flag.)

To take your pet's temperature, insert a lubricated rectal or digital thermometer about one inch into the rectum. Wait one minute, then check the temperature. Anything over 103°F means a fever and a quick trip to the veterinarian. Do not give your pet anything like aspirin unless your vet prescribes it. *Never* give your pet acetaminophen or ibuprofen—it can be dangerous, even fatal.

Although the exam sounds time-consuming, with practice you'll fly through it in about five minutes. Also, you'll enjoy this time with your pet, and your pet will enjoy all the attention. Plus you'll be your veterinarian's most helpful client. Trust yourself as your pet's best observer. If you notice anything that concerns you, see your veterinarian as soon as possible. Such fast action could save your pet from serious discomfort or even death.

Getting to the Root of an Itch

——

Healthy skin is more than a cosmetic concern; it protects your pet's body from disease and injury. It also conserves water, aids in heat regulation and synthesizes vitamin D. Ailing skin cannot optimally perform these vital functions. And skin trouble can be a symptom of internal disease. That's why it's important to deal with skin ailments as soon as they appear.

The first step is to identify the cause. Sometimes that's easy. In winter, for example, indoor heating can cause your pet's skin to dry out. The signs: white flakes (like dandruff) and mild scratching. Treating dry skin is simple.

Supplement your pet's diet with a complete oil. Corn, safflower, peanut and sunflower are some of the oils that contain all the essential fatty acids.

Give your cat about ½ teaspoon at each meal; give your dog 1 to 3 teaspoons at each meal, depending on the animal's size. (More is not better; oils are very fattening.) Pet stores also sell supplements that contain the essential fatty acids.

Shampoo sparingly. Most pets can get by with fewer baths in the winter. When you must bathe your pet, use a shampoo formulated for animals.

Humidify your house. This can benefit your pet's dry skin as well as your own.

If these measures don't clear up the problem, or if your pet has other symptoms such as severe itching, scratching, biting, hair loss, excess oil—even a bad odor—consult your vet. He can perform the skin and blood tests that will identify the root of the problem. Here are some possible offenders.

BACTERIAL DISEASES

Pyodermas (bacterial infections of the skin) may cause pimples, redness and a whitish discharge. Many are itchy. Treatment for pyodermas may involve oral antibiotic therapy and special shampoos.

Acne is one type of pyoderma that is very common on the chins of cats. Treatment: oral antibiotics or local scrubs with antibacterial soap.

Hot spots, also known as acute moist dermatitis, are very common in dogs, especially those allergic to fleabites. These red, moist lesions develop very rapidly as a result of licking, biting or scratching. Short-term treatment with corticosteroid injections is usually prescribed.

FUNGAL DISEASES

Ringworm (*Microsporum canis*) is the most common fungal disease of the skin. Some animals, especially adults, can be carriers. Young animals are more likely to develop skin lesions—usually circular areas of scaling and hair loss. Healing of the lesion occurs from the center outward, causing a ringlike appearance; hence the name. Treatment may include special rinses or shampoos, creams or lotions and oral antifungal medication. Because ringworm can be passed on to people and other animals, it is extremely important that your pet be treated as soon as possible. If you have lesions, too, you should be treated by *your* doctor at the same time.

PARASITIC DISEASES

Fleas are by far the most common skin parasite that plagues our pets. Their bites usually lead to itching and scratching, especially in the rump area. Many dogs and cats are allergic to fleabites and develop a severe generalized skin condition—often red and itchy, sometimes scabby or flaky. If you suspect fleas, look for "flea dirt" in your pet's coat. It looks like coarse pepper but is actually dried blood and flea feces. When wet, it turns red. Treatment consists of eradicating the fleas from the house, the yard and all pets in the household. (See chapter 8 for more on fleas.)

Ticks may cause some local skin irritation, but the diseases they carry are potentially a much more serious problem. Both people and animals can become very ill if bitten by a tick carrying Lyme disease, for example. Prompt removal of ticks with tweezers placed right next to your pet's skin and periodic dipping reduce the chances of infection.

Mites are spiderlike parasites that live on the skin. Two common mites, *Demodex canis* and *Sarcoptes scabiei,* cause mange in dogs. While demodex are found in the hair follicles of almost all dogs, they cause a problem only when immunity flags. Then these mites proliferate and cause hair loss. Most cases remain localized and resolve by themselves. On rare occasions, special veterinary treatment is needed.

Unlike demodex, sarcoptes almost always cause a problem; intense itching, scaling and hair loss, especially around the elbows and ears, are common. Sarcoptes mites are extremely contagious; they spread to other animals and man. Dips, oral medication or injections may be needed to banish them.

Ear mites (*Otodectes synotis*) cause a dark discharge in the ear canal that resembles coffee grounds. Head shaking and scratching and pawing at the ears are common signs of mites (and other ear problems). Insecticidal ear drops clear up this skin problem.

ALLERGIC DISEASES

Unlike humans, who respond to allergies most commonly with respiratory signs, dogs and cats display allergic symptoms most often in the skin—redness, itchiness or scaliness—and sometimes in the gastrointestinal tract. Depending on the cause, removing the offending substance from the diet or the environment, allergy shots or therapy with corticosteroid drugs may be indicated. (For more details, see chapter 9.)

ENDOCRINE DISEASES

Many times endocrine disorders (hormone imbalances) are first noticed because of skin and coat changes—primarily equal amounts of hair loss on both sides of the body. These areas are often darker than normal and are not itchy. Also, the changes do not occur suddenly but take place over a few months.

Hypothyroidism is caused by the decreased production of the hormones of the thyroid gland. It's common in dogs. Their coats become thin, dry and lusterless, and the hair is easily pulled out. Treatment: thyroid supplements.

Hyperadrenocorticism (Cushing's disease) is caused by overproduction of the hormone cortisol by the adrenal glands. Dogs with this condition have thin skin, potbellies and the typical endocrine hair-loss pattern. Therapy with a drug that destroys the abnormal parts of the adrenal glands may be needed. Occasionally, surgery or radiation therapy is recommended.

• 8 •

Strategies for Fighting Fleas

———

C hances are pretty good that you will have to contend with fleas at some time during your pet's life. Not only are these minute, bloodsucking insects the most common pet parasites, they also cause the most common skin disease in dogs and cats. What's more, fleas are not all that particular about whom they take their meals from, so they often bite humans.

As most pet owners know, getting rid of fleas can be difficult, but it's not impossible. The aim is to have a safe and effective program that eliminates fleas from both your pet and its environment and keeps pesticide use to a minimum. You need to detect signs of infestation early and institute your control program as soon as possible.

THE SIGNS OF TROUBLE

If you are finding red, itchy spots around your ankles, it's likely that your dog or cat has fleas. The signs in your pet may vary, though. Most pets do become itchy. The first signs are chewing, licking and scratching, primarily over the lower back and the inside and back of the thighs. The base of the tail is also a favorite place for fleas to bite. And some cats begin to scratch around their necks.

In pets with flea allergies, the itching is severe. The allergic reaction, combined with the excessive licking, scratching and chewing, often leads to hair loss, redness, scabbing and crusting. At this point, the animal is usually very uncomfortable and needs veterinary attention. If the disease goes untreated, it may spread. (See chapter 9 for more information about fleabite hypersensitivity.)

CONFIRMING THE INFESTATION

Once you suspect an infestation, a careful examination for fleas is essential. (It's a good idea to check for fleas on a regular basis as part of routine grooming.)

Run your hand through your pet's coat from head to tail to look for fleas and "flea dirt," which is a combination of dried blood and flea feces that resembles coarse pepper and dissolves into a bloodlike stain when sprinkled on a wet paper towel. You can buy a special flea comb from most pet stores to aid in your flea detection. Fleas are brown, hard insects that are flattened from side to side. Fleas don't have wings, so they can't fly. They can hop remarkably long distances, however, and you may spot them doing just that.

Since tapeworms often result from swallowing infected fleas, the presence of tapeworm segments is also evidence of flea infestation. Check your pet's anus for these ricelike worm segments. You may also see tapeworm segments around the house or in your pet's feces.

If your pet is extremely allergic to fleas, you may not find any actual fleas or flea dirt. Itching along the pet's back and hind legs may be the only sign you'll find.

TAKING CONTROL

Before you attempt to control fleas, it's important to realize that a flea spends most of its life off its animal host. In fact, experts estimate that for every flea on your pet, there are 100 in the animal's environment. So you must kill the fleas living in your house as well. The idea is to use the least toxic products that will still kill the fleas—and to use them properly. In

Antiflea-Product Precautions

A wide variety of antiflea products are on the market. As an informed consumer, you should be aware of their advantages and shortcomings.

Flea shampoos. These usually kill fleas on your pet but do not prevent new fleas from jumping on.

Flea collars and tags. While these may repel fleas, they don't do a good job of killing them. Furthermore, flea collars can cause allergic skin reactions in some animals. (The new electronic flea collars show promise when used with environmental control measures.)

Flea powders. Powders kill fleas and prevent new infestations, but they are messy and do not adhere well to dogs and cats with short coats.

Oral products. These agents work by entering your pet's bloodstream. When a flea bites, it is killed when it ingests the blood. Because these products work only after the flea has already bitten, your pet may still suffer from reactions to the bites. Also, the safety of feeding insecticides to your pet is questionable.

Garlic and brewer's yeast. These nutritional remedies have been advocated in many anecdotal reports. They are harmless and may be worth a try. But the few controlled studies that have been done have failed to demonstrate effectiveness.

this way, you'll use a minimum of insecticides and nip your flea problem in the bud. Be sure to include all your dogs and cats and to take all the measures on the same day.

Here's what I recommend on your pet.

Use a flea dip that contains microencapsulated pyrethrins. Pyrethrins are "natural insecticides" derived from chrysanthemums. Products containing microencapsulated pyrethrins have a delayed release of their active ingredient, so they kill fleas on your pet and continue to kill them for about a week. Pyrethrin sprays are also effective, but most don't last and must be used daily. Pyrethrin products should be used only with your veterinarian's guidance.

Always read the label before applying any product to your pet, even "natural" formulations. Many products are not safe for cats and young animals, and the label says so.

Take these measures in your house.

Use a fogger to kill adult fleas. Look for those that rely on pyrethrins—especially microencapsulated pyrethrins, which, as explained above, have a lasting effect. The best foggers also contain the insect hormone methoprene, which prevents the maturation of flea larvae. Follow package instructions carefully. Or hire a licensed pest-control service and ask them to use a microencapsulated pyrethrin formula.

Vacuum rigorously. Vacuuming is important because insecticides aren't effective against fleas that are in the egg stage. After vacuuming, discard the bag because fleas that hatch inside it can rapidly reinfest your house.

Wash your pet's bedding in hot water. Include any other washable items that your pet regularly comes in contact with.

Take this precaution in your yard.

Spray with a 0.75 to 1 percent solution of malathion. This insecticide is less dangerous than most. Once it dries, malathion is not toxic and does not persist in the environment. So the key to using it safely is to keep kids and pets inside until it dries.

If you live in an area where fleas thrive, such as in the southern states, don't expect one or two treatments to take care of the problem. Outside areas may require insecticide application as frequently as every two weeks.

Very often your pet will continue to scratch for a few weeks after you have effectively eliminated the fleas. That's because your pet is responding to the effects of earlier bites rather than to the presence of the fleas themselves. If you have an allergic pet, it may require anti-inflammatory medication from your veterinarian.

• 9 •

Help for the Allergic Pet

━━━

When humans have allergies, they get sneezes, runny eyes and red, itchy noses. But our pets seldom get these respiratory symptoms when they have allergies. Instead, they tend to get problems of the skin or gastrointestinal tract.

But whatever the symptoms, with the help of your veterinarian you can almost always prevent or minimize them. Here are some good strategies for outmaneuvering some of the more common pet allergies.

FLEABITE HYPERSENSITIVITY

Probably the most common allergic reaction seen in dogs and cats is caused by the saliva of that tiny six-legged insect the flea. Though many animals can be infested with fleas and show no more than an occasional scratch, animals that are allergic need only one fleabite to send them into a frenzy of itching, scratching and chewing.

Depending on where you live, flea allergies may or may not be seasonal. In warmer climates, fleas are present all year long. But in places where there is a winter freeze, fleas seldom become a big problem until late spring through early fall.

Flea allergy is the number one cause of summer itch. Skin changes associated with a flea allergy are variable but usually include itching, redness, scabbing and hair loss. In dogs, the problem areas are usually along the lower back, flank and inside and back of the thighs and on the lower portion of the abdomen. But sometimes the neck, the front of the forearms and the base of the ears are also involved. In cats, the skin sores are usually very crusty. They begin around the neck and lower back and then extend along the cat's entire back.

Because it takes only a few fleabites to cause a reaction in a very allergic animal, fleas may not be visible to you. Most often the fleas, after taking their blood meal and injecting saliva, hop off your pet and onto your carpet or lawn to spend the majority of their lives. Sometimes you may find evidence that fleas have been there in the form of "flea dirt." But even if you don't see fleas or flea dirt on your pet, if the distribution of sores suggests a flea allergy, your pet is probably flea allergic and should be seen by your veterinarian.

The treatment for a flea allergy consists of anti-inflammatory medication (steroids or antihistamines) to reduce the itching, plus stringent flea control.

Good flea control is an absolute necessity, both to treat the problem and to prevent recurrence. Since the flea spends its life not only on your pet but also in your house and yard, all must be treated with some type of insecticide. We all know that insecticides can be extremely toxic to both animals and man, but until other, safer methods of therapy are discovered, we are stuck with them. Some of the newer insecticides, such as the microencapsulated and synergized pyrethrins and the insect growth regulators, are the safest and are very effective. As discussed in chapter 8, your veterinarian can help you set up a flea control program that is best for both you and your pet.

ATOPIC ALLERGIES

When allergic people inhale pollen, dust and molds, they get hay fever and sometimes asthma. They're said to have an atopic, or inhalant, allergy. Dogs and cats can have the same thing. But though some of them can also show hay fever–like symptoms, they most commonly itch.

Classically, atopic dogs rub their faces, muzzles and eyes, scratch their armpits and ears and chew on their feet and legs. Sometimes the itching spreads over their whole bodies, and their skin is red and irritated from self-chewing. With time, skin changes such as hair loss, thickening, scaling, darkening and infection also develop. Cats with atopic skin changes may show crusty sores, hair loss and red, raised areas.

Any time your pet is itching and scratching excessively and seems uncomfortable, it should be seen by your veterinarian. If the clinical signs and history suggest atopic disease, your veterinarian may advise several therapies. When atopic disease is seasonal, it's often treated with a short course of anti-inflammatory medication, such as a corticosteroid. When it becomes a year-round problem, your veterinarian may advise special skin tests or blood tests in order to find out what your pet is allergic to. Then every attempt is made to avoid exposing your pet to those triggers. But because it's rarely possible to prevent all exposure, your veterinarian may recommend a series of hyposensitizing allergy shots.

FOOD ALLERGIES

Allergic reactions to some component of a pet's diet may also lead to itching and a variety of sores almost anywhere on the animal's body. Vomiting and diarrhea may occur simultaneously. Dietary substances that have been shown to produce allergic reactions are numerous and include beef, pork, chicken, cow's milk, horse meat, eggs, wheat, oats, fish and fungal contaminants in drinking water.

The only way to diagnose a food allergy is by putting your pet on a hypoallergenic, or elimination, diet, which consists of only a few foodstuffs to which your pet has had little or no previous exposure. Favorites of veterinary dermatologists are a lamb and rice diet for dogs and a lamb diet for cats. Tap water is replaced with either bottled or spring water. The new diet is fed for a full three weeks, then the pet's skin condition is reevaluated.

If there has been improvement, a food allergy is present. Individual foodstuffs are then added to the restricted diet at five- to ten-day intervals until a balanced diet is achieved or the cause of the allergy is discovered. Diet trials are best done with veterinary supervision, so you can be sure that your pet remains healthy throughout and that the skin condition is monitored objectively.

CONTACT ALLERGIES

Because animals are covered with hair, allergic reactions caused by substances they touch are not as common in pets as they are in us. Dogs, however, do get allergic contact dermatitis from a variety of irritating substances, affecting the hairless parts of their bodies (armpits, lower abdomen, groin, nose, bottom of the feet and between the toes). Itching and redness are the primary clinical signs. Some common contact allergens are plastic food dishes, cleaning agents, detergents, shampoos, grooming products, carpet cleaners, fertilizers and some plants. (Unlike man, dogs and cats do not generally react to poison ivy, however.)

If your veterinarian suspects a contact allergy, she may suggest removing your dog from its usual environment, possibly by boarding the animal, for a short period of time. If there's improvement, the dog will then be challenged with the suspect items one at a time. If an allergy is present, signs usually reappear within 24 hours. Treatment consists of avoiding whatever the dog is allergic to.

GASTROINTESTINAL ALLERGIES

Allergies to food not only may cause skin problems but also may lead to serious gastrointestinal ills. Common causes are preservatives in commercial pet foods, soybeans, wheat and gluten. Although initially causing diarrhea, vomiting and abdominal pain, food allergies often lead to severe chronic diarrhea accompanied by weight loss.

Again, a diagnosis is made by feeding an elimination diet. If there is improvement, other diets may be tried to see if they elicit reactions. There are several gluten-free diets available commercially and some that do not contain preservatives. To prevent recurrence of signs, every attempt should be made to avoid exposure to the offending foodstuff.

There are other specific diarrheas in which an allergic cause is suspected. Two of them in dogs are eosinophilic enteritis and plasmocytic-lymphocytic enteritis. Any chronic diarrhea problem should be checked by your veterinarian. He should be able to tell you if the problem is allergic in nature and if he can help control it.

ALLERGIC BRONCHITIS

Sometimes an inhaled substance, such as tobacco smoke, aerosol spray or cat litter dust, leads to an allergic reaction in the respiratory tract, resulting in a syndrome known as allergic bronchitis. Feline allergic bronchitis is often referred to as feline asthma. Affected animals may cough, wheeze and have trouble breathing. Exercise can aggravate the problem.

If you notice any of these signs in your pet, you should visit your veterinarian immediately. She can diagnose the cause of the problem through physical examination, chest x-rays and sometimes microscopic examination of respiratory secretions. If your veterinarian diagnoses allergic bronchitis, she may prescribe medication that will make breathing easier (bronchodilators) and will help suppress the allergic reaction (corticosteroids). If you suspect that a particular substance is the culprit, it should be removed from the animal's environment as soon as possible.

"INJECTION" ALLERGIES

Like people, dogs and cats occasionally develop a severe allergic reaction, called anaphylaxis, to insect stings, injected drugs, vaccines or blood products. The first sign may be swelling around the eyes, mouth and ears or hives. Sometimes salivation, difficulty breathing, vomiting, diarrhea and abdominal pain follow. If not treated immediately, some animals become uncoordinated, collapse and die. If you notice any of the early signs in your pet, even if you have not seen an insect nearby, take it to your veterinarian immediately. This type of allergic reaction is a true emergency.

Dental Care for Your Pet

Many people believe dental disease is a problem that affects only human beings. On the contrary, dogs and cats suffer from the same kinds of dental disorders, which can be just as painful for them. Also, these ailments can affect the overall health of our pets, just as dental problems can impact on our general well-being.

Luckily, we can prevent most of these diseases in our pets the same way we can in ourselves. Regular cleaning helps keep dental disease from taking hold in the first place. And periodic exams by both you and your veterinarian let you spot problems while they can still be nipped in the bud.

THE CAST OF NASTY CHARACTERS

Here are some of the villains that lurk out there, ready to make life miserable for your pet.

Periodontal disease. This disorder of the gums and other structures that surround the teeth is the most common dental problem of dogs and cats. According to one leading expert in veterinary dentistry, this disease occurs in about 85 percent of all small animals more than six years old.

Periodontal disease is due to plaque and tartar buildup on the teeth. Plaque is a clear, bacteria-laden film that forms on the teeth daily. Unless the plaque is removed by brushing, an encrustation called tartar is formed. The bacteria hiding in tartar first infect the gums, then erode the underlying tissues and bone that support the teeth. Without treatment, the teeth eventually loosen and fall out.

But the trouble may not stop at the teeth. An infected mouth can also affect your pet's health and lead to kidney, liver and heart problems.

In its early stages, periodontal disease involves only the gums and is reversible. The later stages, involving the other structures that surround the teeth, cannot be reversed. To avoid this situation, preventive dental care must be started when your pet is very young.

Small holes in the teeth beneath the gum line (subgingival resorptive lesions). These lesions are found primarily in cats. The holes are often covered with tartar or an overgrowth of the gum tissue and as a result are often overlooked.

Their cause is still somewhat of a mystery. One study related the disease to feeding a pet raw liver. Others suggest a connection between these lesions and chronic gum disease. Commercial cat foods do not seem to be a causative factor, though, since the problems are also found in cats that are wild.

Usually, the affected tooth has to be extracted. If the condition is caught early, though, the tooth might be saved by filling it. To prevent new lesions from forming, a home oral hygiene program should be started immediately.

Broken teeth. This injury is commonly caused by car accidents, by chewing on hard substances or by fighting. If the nerve inside the tooth is exposed, it can be extremely painful. But even if your pet is not in pain, a broken tooth is prone to infection, so it must usually be extracted.

Cancers of the mouth. This problem often occurs in older animals. Check your pet's mouth for growths, lumps or persistent lesions at least once a month. If any of these appear, consult your vet.

THE DENTAL HOME EXAM

Some easy detective work in the form of weekly home exams can ferret out these problems before they turn into health disasters. It's best to start the exams when your pet is very young, so you know what the animal's healthy mouth looks like. That way, your dog or cat will also get used to being examined. Be gentle, move slowly, and use soothing words of praise and special treats.

Here's what to look for.

• Your pet's gums should be pink and glistening. Some animals have dark pigment spots on their gums. This is normal if the areas are not raised and do not grow. Of course, red, swollen, painful, ulcerated or bleeding gums are unhealthy and should be checked by your veterinarian.

• Your pet's teeth should be white. Yellowish brown tartar is a common finding in animals two years of age and older, especially on the back teeth. The accumulation of obvious tartar means your pet's teeth need a professional cleaning.

• A broken tooth warrants a call to the veterinarian.

• Any unusual lumps in your pet's mouth should be seen by your vet right away.

• Bad breath is often an early sign of dental disease.

• Other signs of problems are excessive drooling, reluctance to eat (especially hard food), a swollen jaw or cheek and chronic nasal discharge. If you notice any of these problems, take your pet to your veterinarian.

PREVENTIVE DENTAL CARE

Spotting problems early is only part of keeping your pet's mouth healthy. Taking measures to avoid trouble in the first place is the other. Here's how to prevent tooth trouble.

• Feed your pet dry food or hard biscuits daily. This helps reduce tartar, but it's only one of several steps you should take to prevent dental disease.

• Give your dog a hard rubber toy to "exercise" teeth and gums. It works toward keeping them healthy.

• "Brush" your pet's teeth at least twice a week (although every day is best). When you do it, choose an area where your pet feels comfortable, like a special rug or favorite resting place. Small pets are best done in your lap. Use either a gauze pad or a washcloth to remove plaque and tartar. You could also use a toothpaste made especially for pets (available from your veterinarian). If your pet will not tolerate such cleaning, you may need to have your vet scale the pet's teeth every three months to one year.

• Have the vet do a full dental checkup once a year (it can be part of your pet's regular checkup). If there is a significant amount of tartar buildup, the vet will probably recommend a professional cleaning.

Most animals won't sit still to have their teeth scaled, so it's usually done under general anesthesia. At the same time, your veterinarian can give your pet a more thorough exam, carefully remove all the tartar (even that under the gum line) and extract any severely diseased teeth. If your pet needs a more complicated procedure, such as root canal, you may be referred to a dental specialist.

To some people, a preventive dental health program for pets may seem a little extreme. But follow your own instincts; regular oral hygiene from the beginning can keep your pet healthy, comfortable and sweet smelling, too.

· 11 ·

Raising a Heart-Healthy Pet

The heart is by far the most important muscle in the body—that's true for dogs and cats as well as humans. It's responsible for pumping blood containing oxygen and nutrients to all body organs and for carrying waste products away. And while dogs and cats don't suffer from clogged arteries and high blood pressure the way people do, they commonly get other types of serious heart disease, such as valve and muscle defects and heartworm infestation. Fortunately, many of these problems are not only treatable but preventable.

The first step is to schedule an annual checkup with your veterinarian. When the doctor listens to your pet's heart, he can detect murmurs and abnormalities in rhythm, common forewarnings of heart disease.

Don't rely on yearly exams alone, though. Pets can show signs of serious heart disease that you don't need a stethoscope to detect.

You should be concerned if your dog or cat shows any of the following signs: coughing (especially if uncontrollable and at night), difficult or rapid breathing, relative intolerance to exercise, bluish color to the tongue and gums, apparent fainting or noticeable weight gain (especially in the abdomen). Although these signs can be produced by other ailments as well, they are strongly suggestive of heart disease.

But early detection is only half the battle. If you want to safeguard your pet's heart health, you may need to take some preventive steps.

THE LOW-SODIUM DIET

Although dogs and cats seldom develop problems due to high cholesterol consumption, a high-sodium diet can contribute to heart disease in pets. Most veterinarians agree that animals suffering heart failure improve with sodium restriction. But some vets recommend starting sodium restriction long before any calamity occurs. They may suggest it for pets that have heart problems (such as murmurs) that frequently lead to heart failure. At this time, though, its preventive value is controversial.

It's interesting to note, however, that a dog consuming a commercially available diet ingests approximately 65 milligrams of sodium per pound of body weight per day—about 30 times the amount necessary for normal growth and twice the amount consumed by a person eating a normal Western diet!

There are a few well-balanced, low-sodium commercial diets available for dogs and cats. Most veterinarians carry them. If you and your veterinarian agree that it would be beneficial for your pet to start a sodium-restricted diet, introduce it gradually because low-sodium diets tend to be less palatable than ordinary ones to dogs and cats. Mix it with the old diet, gradually replacing the old with the new. Hand-feeding may help. The complete switch might take four days to four weeks. (See chapter 20 for more about low-sodium diets and their effects on heart disease.)

OBESITY AND YOUR PET'S HEART

There's no direct evidence that obesity contributes to the development of heart disease in dogs and cats. But there's lots of evidence that it does in humans, and there's good reason to believe that the same might hold true for pets. The heart of an obese person or animal has a greater mass to support and therefore has to work harder.

People with heart failure usually improve dramatically with weight reduction, and we presume the same occurs in obese dogs and cats. There are many reducing diets on the market for pets. In fact, many of these are also low in salt. Just be sure to check with your veterinarian for advice before setting up a weight reduction program for your pet. Exercise, although very beneficial for the healthy pet, may be dangerous for one with heart disease.

HEARTFELT AMINO ACID

Cats, unlike other animals, are unable to synthesize adequate amounts of the amino acid taurine in their bodies, so they must get it from their diets. Deficiency of this nutrient can lead to a variety of problems, including heart disease. Research showed that cats with dilated cardiomyopathy, a deadly heart muscle disease, had low levels of taurine in their bloodstreams. When the diets of these cats were supplemented with 250 to 500 milligrams of taurine daily, dramatic improvements resulted.

Strangely, most of the cats studied had been eating commercial cat foods that contained up to four times the amount of taurine recommended by the National Research Council. Why these cats were unable to use the taurine in the food is still unknown. But the report caused most cat food manufacturers to increase the amount of this amino acid in their foods to ensure adequate levels in all cats that eat their products.

Your cat may still be at risk, though, if you are feeding it only dog food, which contains less taurine than cat food (among other crucial differences), or home-cooked meals low in this nutrient. The taurine requirement for cats is about 50 milligrams per day, the amount in three tablespoons of minced clams.

PREVENTING HEARTWORM DISEASE

The heartworm, or *Dirofilaria immitis,* is an extremely common cause of heart disease in dogs (less so in cats). Mosquitoes pick up the worms (called microfilaria at this stage) from the bloodstream of an infected animal. The worms are then injected into other animals when the mosquitoes bite. Once inside, the worms mature and take up residence in the heart. Heartworm infestation causes signs of heart disease: coughing, rapid breathing, exercise intolerance, weight loss and, rarely, sudden death.

The good news is that in dogs, heartworm disease is completely preventable! The first thing you must do is find out if your dog already has heartworms (prevention will be ineffective and can even be dangerous if infection is already present). Your veterinarian can easily check your dog with a simple blood test. If it's positive, treatment should be started immediately.

There are now two excellent preventive agents available. One, diethylcarbamazine, or DEC, is a time-proven product that must be given every day. It's available in liquid and in chewable and nonchewable tablets. The other product, newer and equally safe, is called ivermicten (brand name Heartguard). This has to be given only once a month and is available in both chewable and nonchewable tablets.

No matter which you use, therapy should be started in the spring and continue for at least 60 days after the mosquito season ends.

· 12 ·

Easing Troubled Joints

———

Our pets' joints, just like our own, function to unite bones and aid movement. And like us, when their joints become inflamed or damaged, our pets can become miserable and even be immobilized by the pain. Luckily, they can be put back on their feet.

Joint damage can result from injury (such as a torn ligament), developmental problems (such as hip dysplasia, a genetic disease that causes joint instability), infections (such as Lyme disease), disorders of the immune system (rheumatoid arthritis and lupus) or cancer.

The general term *arthritis* literally means inflammation of the joint and commonly refers to any type of joint degeneration. It may occur for no apparent reason, or it can result from any of the conditions mentioned above. Joint diseases can be inflammatory, with signs of redness and heat, or noninflammatory, as in osteoarthritis, for example.

WHEN JOINTS ACT UP

The first hint of noninflammatory joint disease in dogs and cats is usually a reluctance in the pets to do certain things they readily did before, such as sitting up or jumping up. As it worsens, lameness and stiffness may occur following periods of activity or overexertion. Stiffness may also be obvious after the pet gets up from resting. With continued movement, most animals seem to "warm up" out of their pain. And just as in people, cold and damp weather often makes the signs worse. In cases where joint problems are caused by a disease such as hip dysplasia, lameness or gait abnormalities may precede the signs of joint degeneration by months or years.

Dogs and cats with inflammatory arthritis are not only stiff and lame but also generally "sick." They often seem depressed and have fevers and poor appetites. In many cases, their joints are swollen, red and warm when touched. Dogs suffering with Lyme arthritis are usually lame in one or more joints and are often sick. When the synovial fluid from their joints (a thick fluid lubricating the joints) is examined, it shows inflammatory changes. And when their blood is tested, antibodies to the Lyme disease organism are found. Dogs with rheumatoid arthritis show similar signs but do not have Lyme disease antibodies.

Dogs and cats with joint disease, no matter the cause, tend to be irritable and reclusive. They may snap or bite when approached or touched—especially by children, who often move quickly and unpredictably. These animals are cranky because they're in pain!

RELIEF FOR ACHING JOINTS

If your pet is stiff or lame or shows other signs of joint problems, see your veterinarian. The best treatment for each type is different, so an expert diagnosis is critical.

Your vet will probably recommend x-rays and possibly blood tests, plus analysis of the synovial fluid. Once the diagnosis is made, proper measures can be taken to halt the disease. Your vet will probably recommend a treatment program that includes the following measures.

Medication. Pets with inflammatory arthritis often require antibiotics or cortisone. And any type of arthritis may necessitate analgesic or anti-inflammatory drugs to relieve pain. Buffered aspirin, phenylbutazone and cortisone are often prescribed to control pain in severely affected dogs. (Since cats seldom show signs of joint disease to the same extent as dogs, they rarely need drug therapy.) Don't give your pet over-the-counter arthritis products unless recommended by your vet. Some of these, such as ibuprofen and acetaminophen, can be very dangerous to dogs and cats.

Proper exercise. Pets with noninflammatory arthritis should continue exercising to maintain muscle tone and keep the joints loose. (Of course, strenuous exercise that could result in lameness should be avoided.) A good exercise program includes periods of walking, swimming (the exercise kindest to joints) or playing, interspersed with short periods of rest. Cut back on these activities as the disease becomes more advanced and less exercise is indicated.

Adequate rest periods. It's important to prevent excessive use of damaged joints. Although your dog may be eager to jog with you for five miles, doing so is certain to cause the animal pain the next day.

Weight reduction. Excess weight puts added stress on damaged joints and increases wear and tear. If your animal is too heavy, ask your vet about a good weight reduction program.

Surgery. Your vet may recommend this to relieve pain, regain motion or correct deformities or instabilities. Surgery to remove the head of the femur (thighbone) or to insert an artificial joint often helps pets with hip dysplasia. Surgical correction of a damaged or ruptured ligament in a joint is best done before secondary degenerative changes develop.

Keep an eye out for any signs of joint problems in your pet. Early action might save your dog or cat from months or years of discomfort and limited activity.

• 13 •

Caring for the Older Pet

___

O ur pets are enjoying longer lives than ever before. Thanks to advances in veterinary medicine, a dog's average life span today is 12 years, and a cat's is 14. Most veterinarians consider a dog or a cat past age 7 to be a "senior."

Just as in people, aging brings changes. Some of these changes, such as graying hair, are normal and can't be avoided. Others, however, such as diseases of the heart, kidneys or teeth, may be preventable, or their effects can be minimized.

THE NORMAL CHANGES OF AGING

It's natural for added years to bring the following changes to your pet.

Increased weight. Older dogs and cats often put on extra pounds. This happens for two reasons: They burn calories less efficiently, and their activity levels decrease. If the problem gets out of hand, your pet may need a reducing diet.

Decreased activity. More time sleeping and less exercising are common in older pets. Sometimes the pain of arthritis makes moving difficult, just as it does in humans.

Hearing loss. Maybe your pet isn't responding to commands anymore. That may be due to hearing loss. A deaf pet needs extra care. If your pet can't hear your voice, it also can't hear an approaching car. Keep the animal inside or on a leash when it's outside.

Eye changes. As the normal eye ages, the lens takes on a cloudy appearance. Your older pet may not be able to see as clearly as it once did (losing sight of a squirrel it's chasing, bumping into things when it's in a strange place and the like) but still sees well enough for its needs.

COMMON DISEASES OF THE OLDER PET

Even if your older pet doesn't have obvious health problems, it should visit your veterinarian once or twice a year. Between checkups, watch for symptoms of some of the most common diseases that older animals face.

Kidney disease. As a pet ages, its kidneys often deteriorate. If this happens to your pet, you may notice that it drinks and urinates more than usual.

Heart disease. Coughing and breathing difficulties are symptoms of heart disease. As with humans, the lives of many pets with heart disease can be extended through early diagnosis and treatment. (See chapter 11 for more information about your pet's heart health.)

Dental disease. Bad breath, sore or bleeding gums and yellow or broken teeth point to dental disease. (See chapter 10 for tips on dental care for your pet.)

Arthritis. If your pet is obviously lame or in pain, your veterinarian may prescribe medication. Do not administer over-the-counter preparations, such as aspirin, acetaminophen or ibuprofen, to your dog or cat without consulting your veterinarian. (See chapter 12 for more information on caring for an arthritic pet.)

Diabetes. Diabetic animals usually show excessive thirst, urination and weight loss. Although diabetes is not curable, it can be controlled through daily insulin injections.

Constipation. Older animals are less active, and the nerves and muscles of their intestinal tracts may not function as efficiently, so constipation can be a serious problem. A high-fiber commercial diet is often helpful. Or you can add fiber, such as bran (a teaspoon for a cat, a tablespoon for a dog) or Metamucil (one to four grams for a cat, two to ten grams for a dog, depending on size), to your pet's regular daily diet.

Hyperthyroidism. Caused by an overactive thyroid gland, hyperthyroidism is common in older cats. They may become hyperactive, vomit, have diarrhea and lose weight even though they seem to eat enough. Veterinarians can treat this condition successfully.

HOME CARE FOR OLDER PETS

Attending to an aging pet's medical needs is essential, of course. But other things are important, too—a calm existence, for one.

If possible, try to limit the amount of change in your pet's life. Keep to a routine of feeding and moderate exercise at the same times every day. If you go on vacation, try to get someone to either stay at your home or visit regularly to feed and exercise your pet (instead of boarding your animal at a kennel).

Your older pet's diet should be tailored specially for the animal's years. Cut out excessive protein, phosphorus and sodium because in high amounts they can contribute to kidney and heart disease. Avoid commercial food products labeled "high pro" or "high protein." Also, older pets do best on fewer calories because they use less energy. But don't reduce the amount of food; just choose lower-calorie foods that contain all the necessary nutrients, particularly vitamins A, B_1, B_6, B_{12} and E.

· 14 ·

How to Help
Your Pet Live Longer

—

O f course, the best chance for your pet's long
life starts with good genes and a problem-
free birth. Therefore, no matter where you obtain your
pet—breeder, shelter or pet store—it is best to schedule a
checkup by a veterinarian as soon as possible to make sure
you have a healthy animal to begin with.

During this initial examination, the veterinarian can
detect hereditary or congenital problems, such as a cleft
palate or heart defects, and infectious diseases, such as distemper and parvovirus. Many of these conditions not only
shorten the life of a pet but also decrease quality of life. If
your new pet does not get a clean bill of health, you may
choose to return it or exchange it for a healthy pet.

PROVIDE LIFETIME
PROTECTIVE VETERINARY CARE

Many pet owners believe that regular visits to the veterinarian are no longer necessary once a dog or cat grows up.
That is not so! Annual physical exams, along with annual
booster vaccinations and parasite checks, are among the

best ways to keep your pet healthy. During the annual examination, your veterinarian can detect diseases in the early stages—before they get serious. She will also suggest ways to prevent potential pet problems.

FURNISH A WELL-BALANCED AND CONSISTENT DIET

Veterinarians agree that well-fed dogs and cats stay healthier and live longer than others. Most popular brands of dog and cat foods that you find in your local supermarket fit the bill of a well-balanced diet. To be sure, see that the food is labeled "nutritionally complete and balanced." It doesn't matter whether you choose dry, canned or semi-moist, but try to stay away from generic or supermarket-brand foods. Some of these have proved to be undesirable because the animals absorb the nutrients in them poorly. Also, table scraps are not nutritionally complete and tend to encourage obesity and intestinal problems.

A consistent diet is best for your pet, even though it may seem boring to you. In fact, changing diets often leads to vomiting and diarrhea. Overfeeding, however, is probably the biggest mistake people make when it comes to diet. This leads to obesity, a condition that increases your pet's risk of heart disease, joint disorders and other serious health problems.

EXERCISE YOUR PET REGULARLY

Most pets act younger when they begin regular exercise programs. Dogs regain some of their puppy playfulness, and cats, their kittenish ways. Physically fit pets have fewer health problems and are less likely to be obese.

Exercise also helps reduce behavior problems associated with confinement, such as destruction of furnishings and overactivity. Try to exercise your pet at least 15 to 30 minutes daily. Walking, jogging and ball playing are good ways to exercise your dog. Chasing toys on strings or Ping-Pong balls keeps most cats in good shape.

PLAY WITH YOUR PET

This advice certainly makes sense, but many people don't know how to do it. Pleasurable as it may be for both of you, it is not enough to just stroke and cuddle with your pet. Dogs and cats are inherently very playful and social animals. When your dog crouches in front of you with its hind end up in the air and its tail wagging, or when your cat comes darting at you from behind the couch, it wants to play! Most dogs enjoy playing fetch and chasing objects. Cats like to chase lightweight balls or toys on strings.

Not only does play help keep your pet in shape, it also improves the quality of your pet's life. Daily play also enhances that all-important bond between you and your pet.

KEEP UP WITH GROOMING

A daily grooming session helps keep your pet's coat in peak condition and provides a good opportunity for a home physical examination. As you brush your pet's coat, check the ears for any discharge or redness, search for external parasites such as fleas and ticks, see if nails need trimming, and wipe your pet's teeth with a gauze pad or washcloth to remove plaque buildup.

This session should also be used to touch and massage your pet all over. Not only will it make both of you feel good, it will also enable you to detect any changes, such as growths, that your veterinarian should be made aware of. This is also a good time to check for obesity by feeling for your pet's ribs. If you can't feel them, cut out the extra dog biscuits, add an extra walk—or do both.

NEUTER YOUR PET

Neutering (spaying in females and castration in males) prevents your dog or cat from adding to the pet overpopulation problem by reproducing. But even if you confine your pet and prevent any contact with other animals, neutering is still desirable for its numerous benefits to both health (protection from dangerous uterine infections and tumors plus a drastically reduced incidence of breast cancer in females) and behavior (a reduced incidence of roaming, mounting, urine marking and aggression in males). For more about the advantages of having your pet fixed, see chapter 4.

TRAIN YOUR PET
TO BE A GOOD COMPANION

Behavior problems are among the most common reasons that people give their pets to shelters. Proper training early in life can prevent such trouble. For dogs, training not only teaches the animal basic commands but also teaches you how to communicate effectively. The routine commands— "sit," "stay" and "come"—may someday stop your dog from running into a busy street or other dangerous actions. Training your dog to walk on a leash under control is the best way to prevent accidents and fights.

Our feline friends do not do very well in routine obedience classes. However, some training is also very important for them in the interest of preventing behavior problems, such as scratching furniture and jumping on counters and tables to eat your food. Use spray from a plant mister as a deterrent, and provide a scratching post and sufficient cat food plus cat treats as a distraction.

PRACTICE GOOD ANIMAL
CONTROL AND MANAGEMENT

If your dog or cat never has a chance to get into a dangerous situation, then it won't get hurt. A pet that roams can raid garbage cans, sample rat poison or snail bait and drink antifreeze, all which can make it very sick.

Training your dog to stay in your yard is practically impossible without a fence. Fences, like leashes, protect dogs and people. They are not cruel.

Cats that spend their entire lives indoors live longer than cats that roam freely outside. The indoor life protects cats from cars, fights and diseases. If you do decide to keep your cat indoors, start at the kitten stage. You can train your cat to walk on a leash—but don't expect it to heel perfectly. It is best to let your cat walk you.

DEMONSTRATE AFFECTION
AT EVERY OPPORTUNITY

We know that petting makes a dog's heart rate go down. Most owners can describe the relaxed look of pleasure on their pets' faces during petting. Although it is not scientifically documented, we know our pets benefit from being petted, paid attention to and loved.

Spend quality time with your pet. Play, exercise, good food, training and petting all demonstrate your love.

• 15 •

Protecting Your Family from Pet-Transmitted Illnesses

Over the past decade, studies have confirmed what many pet owners consider obvious: Pets are good for our health. Pets provide a source of constancy in our lives, give us companionship, make us laugh and even exercise with us. These attributes enhance the quality of our lives and foster good health.

But pets do occasionally get diseases that can be passed on to people. Should this be a cause for concern? For the most part, no. For one thing, such diseases are rare; for another, there are many ways to prevent their transmission to people.

The greatest fear seems to have been generated by unfounded claims. A while back, for example, researchers uncovered a statistical association between the occurrence of canine distemper in small dogs and multiple sclerosis (MS) in their owners. Subsequent studies have failed to duplicate those results. (Dogs should be immunized against distemper anyway, for their own welfare.) Also, because of similarities between the AIDS virus and the feline leukemia virus (FeLV), some people have been concerned about catching AIDS from FeLV-infected cats. To date, there is no evidence to suggest that FeLV causes AIDS or any other disease in human beings.

CLAMP DOWN ON BITES

In America, animal bites are the most common human health hazard associated with pets. In one study, the annual rate of serious bites was estimated at 1 in every 170 people. Children are the most frequent victims, probably because their unthinking behavior often provokes bites.

Most pet owners are well aware of the danger of contracting rabies from animal bites. For your own protection, make sure that your pet's vaccinations are up to date.

Bites can be prevented through proper education. This is especially true for children. Teach them to move slowly around animals and to be gentle when petting and playing with them. Tell them not to tamper with pets' food dishes or to disturb them when they're sleeping. It's unwise to approach stray animals. And when greeting an owned animal, always ask the owner if the dog or cat is friendly before petting it. If a dog chases you, it's usually best not to run. Some animals might see a running child as escaping prey, resulting in bites and even more dire consequences.

GOOD PET HYGIENE

There are several bacterial infections that can be passed from animals to their human companions. Salmonella and campylobacter are two examples. Both dogs and cats (plus other animals such as turtles, chickens and horses) can harbor such bacteria, even though they often show no symptoms. In people, though, infection can cause stomach upset and diarrhea. As with many gastrointestinal diseases, the key to prevention is proper hygiene.

Young children are most commonly affected, probably because of their careless sanitary habits. Children should be taught not to handle their pets' food or feces. Of course, washing hands after contact with animals is good sense for everyone, especially before eating (these bacteria are transmitted orally). If your pet has diarrhea, separate the animal from your family and contact your veterinarian.

Streptococcus, the bacterium that causes strep throat, is occasionally contracted from a household dog. If a member of your family is having a problem with chronic strep throat, consider having your dog tested.

There are other bacterial diseases that pets can pass to humans, but they're much less common.

Cat scratch disease, or cat scratch fever, is not serious and usually goes away without treatment. You can prevent it by avoiding cat scratches and bites and by not handling stray or unknown cats.

Leptospirosis, a disease that affects the kidneys and liver, can be passed on from dogs to people. The best method of prevention is annual vaccination of your dog with the basic combination vaccine, so the animal cannot become infected.

Yersinia pestis, the bacterium responsible for bubonic plague, is a problem in the southwestern United States. The disease is usually transmitted to man by fleas but can be spread directly from animals. Free-roaming pets are considered an important source of this disease, so keeping pets properly restricted is essential. Flea eradication also helps prevent infection.

Roundworm (toxocara), hookworm (ancylostoma) and *Giardia lamblia* are examples of some intestinal parasites of dogs and cats that can be transmitted to humans through feces. Again, for all intestinal parasites, good hygiene is key.

To prevent roundworm, a veterinarian should examine your pet's stool every six months for parasites. Most newborn puppies and kittens are infected with roundworms by their mothers and should begin treatment as soon as possible.

Toxoplasma gondii, another parasite, is of special concern to pregnant women because it can cause birth defects. Usually, the disease is acquired by eating raw or undercooked meat or through contact with infected cat feces. (The cats become infected by eating small rodents or birds.)

Pregnant women should wash their hands thoroughly after handling cats and should always wear gloves when gardening, lest they come into contact with cat feces in the ground. Cat litter should be changed daily by a family member not at risk, and the litter box should be rinsed regularly in scalding water for five minutes. Pet cats should be fed adequate amounts of food and restricted from hunting, if possible. All meat eaten by pets as well as humans should be cooked thoroughly, and hands, cutting boards, sinks and utensils should be washed with soap and water after contact with raw meat.

Certain skin diseases can also be carried by pets. Ringworm, a fungal infection, is an extremely common one. Fleas, scabies and other mites may also cause mild skin symptoms in people. All these can be prevented by eliminating the infection on your pet.

Remember, the most effective key to self-protection usually lies in simple preventive measures.

· 16 ·

The Home Pet Pharmacy

When illness causes your pet pain and discomfort, you may be tempted to look to your medicine chest to provide relief. But not all the over-the-counter (OTC) drugs you keep on hand for yourself are safe for pets. And the ones that are safe need to be used in special pet-size doses. Also, eliminating your pet's symptoms can make you think you've cured the disease when in fact the problem lies deeper. Still, you might see no need to seek further treatment—and that can be very dangerous. Obviously, it's best to check with your vet before giving any OTC to your pet.

With your vet's supervision, you can use several OTCs to aid your pet—even some usually reserved for people.

PAIN AND FEVER REMEDIES TO TRY

Aspirin, only in its buffered forms, is the drug of choice for fever and pain in dogs and cats. (Acetaminophen and ibuprofen, although safe for humans, should not be used for pets. In fact, acetaminophen can be fatal to cats.)

If you think your dog or cat has a fever, confirm it first by taking the animal's temperature rectally. If the reading is over 103°F, your pet has a fever that merits professional attention. A lower-grade fever can help your pet fight off disease. A very high fever (over 104° in dogs or 106° in cats) can be dangerous. In that case, with your vet's okay, you can give aspirin, but your pet should see the vet right away.

A safe dose of aspirin for dogs is one-fourth of a five-grain (regular strength) tablet for every ten pounds your dog weighs, twice daily. Cats metabolize aspirin at a slower rate and need only one-fourth of a tablet or one baby aspirin every three days. Higher doses are dangerous.

Aspirin is also useful for arthritis and muscle injuries in pets. If your pet is limping but otherwise healthy, your vet may okay using aspirin as a temporary measure, provided you keep your pet calm and see the vet if the animal is not better in two days. Prolonged pain or limping can signal a nerve or disk problem. In such a case, pain would actually protect the animal by keeping it from jumping around and causing more damage.

MEDICATION FOR COUGH, COLD AND ALLERGY

The antihistamines chlorpheniramine and diphenhydramine are sometimes helpful for treating pets with skin allergies or with runny noses caused by upper respiratory infections. These medications are appropriate only if the condition is mild and the pet has no fever and hasn't stopped eating. Check with your vet.

The chlorpheniramine dosage for a cat or small dog is 1 to 2 milligrams (one-fourth to one-half of a tablet) twice a day; for a large dog, 2 to 4 milligrams two or three times a day, depending on weight. The diphenhydramine dosage for a large dog is 25 to 50 milligrams (one to two capsules) two or three times a day, again depending on weight. Nonprescription diphenhydramine isn't recommended for cats or small dogs because the capsule dosages are too large.

In most cases, the cough suppressant dextromethorphan is safe and effective as a treatment for dogs that have kennel cough, an upper respiratory infection. But bronchitis, pneumonia and a collapsing trachea (windpipe), which have similar symptoms and require medical attention from a veterinarian, are easily mistaken for kennel cough, so see your vet before using medication. Dextromethorphan won't help a pet that is coughing due to heart failure.

Cats can develop coughs due to hair balls, pneumonia or asthma, but dextromethorphan will not help any of these. See your vet before giving any cough suppressant.

DIARRHEA REMEDIES TO CONSIDER

The best treatment for diarrhea that occurs suddenly and is not accompanied by any other symptoms is intestinal rest. Therefore, if your pet develops simple diarrhea, withhold food (not water) for 24 hours. Introduce a bland diet (boiled hamburger or chicken and rice) for two to three days, then gradually mix in the regular diet.

If your vet approves, you can use bismuth subsalicylate to help curb diarrhea, but you must still withhold food. Call the vet back if the problem does not clear up within 24 hours. A safe dose for dogs is one teaspoon for every ten pounds of body weight three or four times daily (or until the stool is normal), but for no more than three days. Cats metabolize this drug slowly, just like its relative, aspirin. If your vet agrees, you may use one teaspoon for every ten pounds of body weight three times a day, for no more than three days.

LAXATIVE OPTIONS OPEN TO YOU

Pet owners often misdiagnose constipation. Surprisingly, very often a pet straining due to diarrhea is thought to be constipated, and a pet so severely constipated that it can only pass fluids is thought to have diarrhea. If your pet has been eating normally but not producing feces for one to two days, it is probably constipated, and you should have your vet examine the animal. Constipation could be a symptom of something serious.

If a diagnosis of simple constipation is confirmed, most veterinarians will prescribe a laxative and a bulk former (psyllium), a stool softener (docusate) or a lubricant (mineral oil or petroleum jelly). Of course, plain old wheat bran is a bulk-forming laxative that works, too. (Never use mineral oil for cats—it can accidentally be inhaled and cause fatal pneumonia.)

Lubricants and bulk-forming laxatives may also be used for cats with hair ball problems, although a regular dose of ½ teaspoon of bran mixed into your cat's food may help hair balls pass through the animal's system. Several kinds of petrolatum-based laxatives available in the pet section of the supermarket are commonly used for hair balls as well.

TOPICAL DRUGS CAN BE EFFECTIVE

Rub-on drugs are rarely given because pets have a tendency to lick them off. Even so, products that contain cortisone are sometimes recommended for hot spots or flea rashes. If your vet concurs, you can apply a dab of the stuff three times a day. If the problem persists after two days, take the animal to your vet. If your vet determines that your cat or dog has a minor skin infection, you may use antibiotic ointments such as neomycin, bacitracin and polymyxin. A dab two or three times a day is enough. Don't use these on a large patch of skin or a bleeding wound, however. And see your veterinarian if there's no improvement in two days.

Experts agree that some over-the-counter drugs are relatively safe and inexpensive aids in treating sick pets under certain circumstances. But caring pet owners will wisely avoid the chance of serious error by consulting with a veterinarian before using any of these remedies to relieve an ailing animal.

PART • THREE

Feeding Your Pet Right

• 17 •

Should Your Pet Take Vitamins?

———

Many of us supplement our daily diets with vitamin pills. But should we do the same for our pets? Many commercial pet foods are labeled "nutritionally complete," yet pet stores and some veterinarians sell several kinds of special vitamin preparations. Is it any wonder that pet owners are confused?

VITAMINS AND ANIMALS

Vitamins are vitally important to a wide variety of physiological processes in animals, just as they are in people. Of the fat-soluble vitamins (A, D, E and K), all except vitamin K are necessary in the diets of healthy animals. (Vitamin K is normally produced by the animal's own intestinal bacteria.) Any excess of a fat-soluble vitamin is stored in the body rather than excreted, so overdoses of these vitamins (especially A and D) are much more common than deficiencies in pets.

Of the water-soluble vitamins (the B vitamins and C), all the B vitamins are necessary in the diets of healthy animals, though some are also produced by bacteria in the intestinal tract. Unlike primates, dogs and cats do not need to rely on their diets for vitamin C. They produce enough of this nutrient in their livers.

A DIET FOR ALL SEASONS

Your pet's nutritional needs change in response to its lifestyle and stage of life. For instance, growing puppies and kittens require different levels of vitamins than older pets.

Special Cases

In some situations, even a pet fed a balanced diet could become deficient in an individual vitamin. If you suspect one of the following deficiencies in your pet, see your veterinarian.

Vitamin A. Conditions that cause fat malabsorption can cause vitamin A deficiency. So can chronic administration of mineral oil. Some symptoms: thickened, dry, scaly skin and night blindness.

B vitamins. Deficiencies of the B vitamins can occur when loss of water from the body is increased, as with frequent urination or diarrhea. Chronic administration of antibiotics can also deplete the B vitamins.

Vitamin C. No signs of deficiency have been reported.

Vitamin D. Fat malabsorption and kidney disease can cause a deficiency of this vitamin.

Vitamin E. Malabsorption syndromes and feeding your pet animal fats to produce a shiny coat can lead to vitamin E deficiency. Muscle weakness and reproductive failure are some symptoms.

Vitamin K. Deficiencies may occur with fat malabsorption, liver disease, chronic use of mineral oil or of antibiotics and from ingestion of vitamin K antagonists such as warfarin, a common type of mouse and rat poison.

Unfortunately, information on optimal requirements rather than minimum requirements is lacking. However, pet food manufacturers must provide adequate nutrients for all stages and styles of an animal's life, so most commercial pet foods already exceed most nutritional requirements for the average pet. It's also reassuring to know that foods labeled "complete" and "balanced," as determined by AAFCO (Association of American Feed Control Officials) protocol testing, have been proven nutritionally adequate.

So if you are feeding your pet a good commercial diet and the animal seems healthy, it is unlikely that vitamin supplementation is needed. In fact, supplementing a good diet may result in nutrient excesses as well as secondary deficiencies (when too much of one vitamin or mineral raises the requirement for another).

There may be exceptions, however. One study did show that some dogs fed canned dog food exclusively (especially if the food had a high meat content) developed vitamin E deficiencies. While this is unproven, it's a good idea to hedge your bets by feeding your dog both dry and canned food. Another caution: If you are using a "natural" brand of dry pet food that does not contain preservatives, be aware that it may lose nutritional value over time. Make sure that such a product is not more than six months old when you feed it to your pet.

A BOOST FOR HOME COOKING

Feeding your pet primarily "from the kitchen" presents its own set of nutritional pitfalls. A few examples:

- Dogs fed only meat do not get the proper amount of the mineral calcium and can end up with rickets (if young) or osteoporosis (if older).
- Cats fed primarily raw fish can develop thiamine deficiencies because certain types of fish (smelt, bullhead, herring, catfish, carp and others) contain an enzyme that destroys thiamine. Cooking destroys the enzyme.

• Dogs and cats fed too many raw eggs can develop deficiencies of biotin, one of the B vitamins, by a similar process.

• Cats kept on vegetarian diets usually become vitamin A deficient.

If you worry that your pet's home-cooked meals are not nutritionally complete, consult with your veterinarian. Together you can determine whether your pet might benefit from taking supplements. Or better yet, ask your vet to design a balanced home-cooked diet—one that includes a variety of meats, grains and vegetables—and follow it.

· 18 ·

Feeding Finicky Eaters

In pets, as in people, a hearty appetite is a sign of good health. So even though obesity is common among household pets, pet owners rarely complain that their pets overeat. The more common complaint is that the pet has grown finicky. Sometimes the pet is more than finicky—it may have a loss of appetite that signals an underlying health problem. Anorexia, the technical term for a lack of appetite, can indicate any one of a number of conditions, ranging from dental disease or intestinal disturbances to respiratory problems or cancer.

So how can you tell whether your pet is ill or simply finicky? Unlike pets with finicky appetites, sick pets show other signs of disease, such as obvious weight loss and dull eyes and coat.

In general, healthy dogs and cats are naturally very good eaters. Most pets that become picky eaters do so later in life because we unwittingly teach them to be finicky.

If your pet looks healthy but just won't eat as enthusiastically as the dogs and cats in pet food commercials, here are some hints to ensure a healthier appetite.

GETTING KITTY BACK TO NATURE

Cats are true carnivores (meat eaters). Carnivores tend to eat infrequent meals consistently high in water, protein and fat. In other words, cats are not natural nibblers or snackers. Feeding your cat twice daily rather than leaving food out all day could help lure your cat back to the food bowl.

Also, cats do well on a consistent diet of highly nutritious mixtures of different meats. Finicky cats that eat only one type of food have usually been taught to do so by their owners. A kitten fed only liver, for instance, might get hooked on that food for life, even though an exclusive liver diet is not nutritionally adequate.

To prevent the development of an addiction, nutritionists recommend that single foods or cat foods consisting primarily of single foods (like some canned gourmet cat foods) should compose no more than 25 percent of a cat's diet. If your cat is already addicted to a single food, try to incorporate variety gradually. Each day, combine proportionately less of the habit-forming food with the mixed ration. No matter what, do not give in and go back to the old diet. You may find in a matter of weeks that your cat has unlearned its finicky attitude.

Cats have been shown to prefer food served at 78° to 103°F. Warming your cat's food in the microwave before serving it often aids in curing feline pickiness.

TEMPTING THE CANINE APPETITE

Dogs are not true carnivores and eat more types of foods than cats do. But they, too, prefer infrequent meals. They are attracted not only to protein and fat but also to sweet foods, which are not good for them. (Most cats ignore sweets.) Although some dogs act as though they are competing for their food and would rather eat alone, most dogs are naturally very social and prefer to eat when people or other dogs are around. Try gently praising and petting a finicky dog while it eats to encourage it to finish. Don't get your dog used to being hand-fed, since this is a luxury that many dogs find hard to give up.

Very often dogs are drawn to foods that have strong odors, such as canned foods. Sometimes warming the food whets a dog's appetite. You can season your dog's food, too. Dogs are attracted to flavors such as garlic and meat broth.

Dogs are all too easily taught to be finicky. Those that are frequently fed table scraps often hold out for something better than dog food. A home-cooked meal may seem more delicious for your pet, but it is very likely unbalanced. Besides, table scraps often cause gastrointestinal upset. If you can't resist cooking for your dog, try to make "people food" no more than 10 percent of your dog's diet, and make sure you ask your veterinarian for guidance in formulating a well-balanced diet.

If you want to unspoil an already spoiled dog, gradually introduce a well-balanced, palatable dog food over a period of a few weeks. And don't give up! If you do, you will only train your dog to be even more finicky. (Don't worry, your dog will not starve itself.)

Make sure you are not just trying to plump up your pet to meet your personal ideal of what looks healthy. Also, many people use food as a way to demonstrate their love for their pets. Often they "love" their pets so much that the animals end up overweight and unhealthy. If you really love your pet, don't overfeed it. Instead, show your love through play, exercise and a well-balanced diet.

· 19 ·

Is Your Pet Too Plump?

O besity is a problem of epidemic proportions among humans, but did you know that it's also the number one nutrition-related disease among pets? Research shows that in the United States, up to 40 percent of dogs and 10 percent of cats carry more weight than is healthy for them.

Animals become fat for the same reason people do: They eat more calories than their bodies need. Very rarely, endocrine imbalance, such as hypothyroidism, triggers obesity. But as a rule, the excess weight is a result of eating too much and exercising too little.

It's not that people don't care about their pets; on the contrary, they're overindulgent. We know our pets love food, and so we offer snacks. It's hard to resist those big, sad eyes at the dinner table! Or we pamper them with rich, home-cooked meals. Research shows that pets fed home-made meals, table scraps and treats have a higher incidence of obesity than those fed only commercial foods.

Unfortunately, obesity has serious consequences. The extra weight can lead to or worsen a variety of medical

Recipes for Pet-Slimming Diets

For Dogs

¼ pound lean ground beef
½ cup uncreamed cottage cheese
2 cups carrots, cooked
2 cups green beans, cooked
1½ teaspoons nutritional bonemeal (available in pet stores, pharmacies or health-food stores)
  Balanced vitamin and trace mineral supplement (available in pet stores; follow manufacturer's instructions for dosage)

Cook the beef, drain the fat, and cool. Add the cottage cheese, carrots, beans, bonemeal and supplement. Mix well.

Yield: 1¾ pounds

Dog's Ideal Weight (lb.)	Quantity of Recipe to Feed Daily (lb.)
5	⅓
10	⅔
20	1
40	2
60	2½
80 or more	3½

For Cats

1¼ pounds liver, cooked and ground
1 cup cooked rice
1 teaspoon vegetable oil
1 teaspoon calcium carbonate (crushed Tums)
  Balanced vitamin and trace mineral supplement (available in pet stores; follow manufacturer's instructions for dosage)

Combine the liver, rice, oil, calcium carbonate and supplement. Mix well.

Yield: 1¾ pounds

Cat's Ideal Weight (lb.)	Quantity of Recipe to Feed Daily (lb.)
5	½
10 or more	⅔

problems—skin disorders, arthritis, reproductive problems, diabetes, heart disease and more. Overweight tends to make a pet lethargic and shorten its life span.

FEEL FOR THE FAT

There's no mistaking the shape of a really obese animal. But if the problem is not severe, owners may be completely unaware that their pets carry extra fat until it causes health problems. The easiest way to evaluate your pet's condition is by looking and feeling. A trim dog or cat is lean and firm. Your pet should not have a sagging abdomen. A dog should have a clearly defined waistline behind the rib cage. Feel the ribs along the underside of the trunk to detect extra fat. You should be able to feel (but not see) each rib. If you can't feel the ribs at all, your dog or cat is dangerously obese.

A WEIGHT LOSS PROGRAM

If you suspect a weight problem, take your pet to a veterinarian. Your vet will be able to rule out serious diseases that resemble obesity (such as heart disease, which causes fluid accumulation). Then you and your veterinarian can design a reducing program.

Calories can be cut in one of two ways: either by feeding your pet less of the diet it eats now or by feeding a special low-caloric, high-fiber diet, which is usually more effective. These diets are available commercially, or you can cook them up yourself, following a prescribed recipe. (See "Recipes for Pet-Slimming Diets" on the opposite page.)

Divide the total daily allocation of food into two or three small meals. Your pet won't feel so hungry and will be less likely to beg for food. Try not to provide snacks. If you can't resist, however, offer low-calorie snacks, such as carrots or green beans.

An exercise program is also an essential part of an effective weight loss regimen. A dog needs to work up to 15- to 30-minute brisk daily walks. Be sure to slow down or stop if either you or your pet becomes fatigued. As your dog's condition improves, you may be able to take it jogging or play fetch. For cats, string chasing for 15 to 30 minutes a day provides a good workout.

MAINTAINING A HEALTHY WEIGHT

On a diet, a small dog can be expected to lose about one pound per week, and a large dog, about three pounds per week. Cats can lose about a half pound each week.

Weigh your pet every few days and keep a written record of its progress. Once the goal has been reached, change to a weight maintenance diet and continue to weigh your pet on a regular basis.

· 20 ·

Healing Diets

We have numerous specific diets for fighting disease in humans—why can't we have diets to battle disorders in our pets? We can. Just as in people, the proper combination of nutrients can help slow the progression of some pet diseases and, in some cases, prevent them from occurring at all.

HEART DISEASE

Although dogs and cats don't get atherosclerosis and heart attacks as humans do (despite high consumption of cholesterol), they often get other kinds of heart disease.

Heart problems are among the most common medical disorders in dogs and cats. In dogs, valve disease, muscle disease and heartworm infestation can often lead to heart disease. In cats, a defect in the heart muscle is the most common heart problem.

When afflicted with one of these problems, animals often go into heart failure, a dangerous but usually gradual process of deterioration. This occurs when the heart is too weak to do its job of pumping blood throughout the body. When this happens, the blood backs up and causes fluid to leak out of the blood vessels.

If the left side of the heart is defective, fluid accumulates in the lungs, which causes coughing, difficulty breathing, exercise intolerance and shortness of breath. When the right side of the heart is affected, the fluid leaks into the abdomen and other parts of the body. This makes the animal look as though it has gained weight, but usually only in the belly.

A sodium-restricted diet is highly recommended. Reduction of sodium intake decreases pressure buildup in the heart, reduces fluid retention and aids in excretion of water by the kidneys. For dogs, a sodium-restricted diet should contain 6.5 milligrams of sodium per pound of body weight for severe restriction and 15 milligrams per pound for mild restriction.

When sodium restriction is instituted, it's very important that the diet remain nutritionally balanced in all other respects. The commercial low-sodium diets ensure this. However, you may choose to prepare a home-cooked diet for your pet with heart failure. If you do, check with your veterinarian first, so you know that the diet you're planning is balanced. And remember, your tap water may be a source of sodium, too—if the sodium level is over 150 ppm (parts per million), use bottled water.

It's unknown whether feeding a low-sodium diet before signs of congestive heart failure occur actually delays the onset of the disease. Many veterinarians do, however, recommend a low-sodium diet if they feel that congestive heart failure is likely.

If your pet is in heart failure, your veterinarian will prescribe medication, such as diuretics. If you follow your doctor's recommendations and feed a low-sodium diet, your pet has the best chance of living a relatively normal life.

KIDNEY DISEASE

Kidney disease is another common medical problem of dogs and cats, especially in older pets. Although kidney disease may be caused by a variety of factors, clinical signs and progress of the disease may be minimized through diet in many cases.

When the kidneys are damaged by disease, toxic waste materials (especially the by products of protein metabolism) begin to accumulate in the bloodstream, causing diminished appetite, bad breath, mouth ulcers, vomiting and diarrhea, which is often black in color. Thirst and urination often increase, but in some cases urination decreases or ceases altogether.

If you notice any of these signs in your pet, the animal should be seen by your veterinarian immediately. Depending upon the cause of the kidney disease, your doctor may prescribe both medication and a special diet.

The diet most commonly prescribed in such cases is lower in protein than ordinary diets but contains sufficient high-quality protein and calories to maintain life. The goal of dietary management of kidney failure is to reduce the work load on the damaged kidney while supplying the nutritional needs of the animal. Feeding a reduced-protein diet has been shown to lessen both further damage and the clinical signs of disease. The current recommendation for dogs in kidney failure is to feed 1 to 1.1 grams of high biological value protein (cooked egg protein) per pound of body weight per day. For cats, which have a greater normal protein requirement than dogs, the recommendation is 1.5 to 1.6 grams of high-quality protein per pound of body weight daily. It is also recommended that you ensure an intake of 32 to 50 calories per pound daily for dogs and 32 to 36 calories per pound daily for cats.

There are commercial diets specifically formulated for dogs and cats with kidney disorders, but if you prefer to prepare your pet's diet yourself, ask your veterinarian for advice.

There is no conclusive evidence to show that protein restriction does anything to prevent kidney damage in a healthy pet. But since kidney disease is so common in older pets and often goes undetected, keeping pets over seven years of age on mildly protein-restricted diets (such as commercial diets sold for geriatric pets) is a good idea.

GASTROINTESTINAL PROBLEMS

Vomiting, diarrhea, flatulence and constipation—whether of sudden onset or of a chronic nature—can often be prevented or treated with the proper diet. Like diseases of all organ systems, gastrointestinal problems can have a variety of causes. These include infection (viral or bacterial), parasites (worms), toxins (from spoiled food, for example), allergies, tumors, sudden change to a new diet and ingestion of nonfood materials (such as dirt and sticks).

For simple and sudden vomiting or diarrhea when your pet does not look or feel sick, the best initial step is to withhold food for 24 hours. If your pet does not look well, or if the problems persist for more than one day, however, see your veterinarian.

Depending on the cause of your pet's symptoms, your veterinarian will most likely prescribe medication and some sort of diet change. If the problem involves the stomach or small intestine, a bland, easily digested diet, low in fat and fiber and high in carbohydrates, is usually recommended. For problems of the large intestine, a diet higher in fiber is best. If your veterinarian suspects that your pet's signs are due to an allergy, he may advise a hypoallergenic diet, usually lamb-based (since lamb is less allergenic than fish, chicken, beef or pork). See chapter 9 for more information about food allergies in pets.

Probably the most common causes of acute gastrointestinal problems in pets are diet-related—such as introducing a new food too rapidly, feeding home-cooked foods or the pet's eating from the garbage container. To prevent these simple problems, remember to introduce a new diet gradually over a period of a week. If you're supplementing your pet's diet with table scraps, stick to the low-fat, blander ones (such as skinless chicken and vegetables). And keep a strong, pet-proof lid on your garbage!

FELINE UROLOGIC SYNDROME (FUS)

Many cats develop this potentially fatal syndrome sometime in their lives. Early signs of FUS include straining to urinate, urinating in unusual locations (such as bathtubs and sinks) and bloody and gritty urine. In severe cases, cats may lose their appetites, vomit and be unable to urinate. These symptoms are caused by an inflammation of the lower urinary tract, combined with the formation of crystals or stones in the bladder and urethra.

Studies have shown that several dietary factors play a role in the development of urinary crystals and stones. FUS is less likely to occur in cats fed diets that are low in magnesium and that promote acidic urine. Many commercial foods now available are specially formulated to prevent and treat FUS. If you feed your cat table scraps, avoid fish, cheese and bones. And do not give your cat vitamin/mineral supplements. Most important, be sure your pet drinks plenty of water, which helps flush the urinary tract. Some vets recommend salting the cat's food to stimulate thirst and promote urination.

If you suspect FUS in your cat, see your veterinarian immediately.

CANINE STONES

Like cats, dogs are also afflicted with stones of the lower urinary tract. Signs of stones are very similar to those in cats: frequent urination (often in odd places), bloody urine, straining to urinate, loss of appetite, vomiting and abdominal pain. Some dogs, however, may have bladder stones and exhibit no signs whatsoever. There are diets designed specifically to dissolve canine bladder stones. If you notice any of these signs in your dog, see your veterinarian immediately—because if your dog is unable to urinate, the situation may become life-threatening.

STARTING THE NEW DIET

Whenever you start feeding a new diet, it's extremely important to follow your veterinarian's instructions strictly. Some diets that help one disease may be harmful if your pet also has another illness.

Unless your veterinarian advises otherwise, a new diet should be introduced gradually. Mix it with the old diet, gradually eliminating the old. If your pet just won't accept the new diet, try warming it, hand-feeding for a few days or moistening it with water.

Some pets are wary of eating unfamiliar foods. Don't leave it to chance when a diet change is vital to your pet's health. Your patience and persistence in getting your pet to eat could save its life.

PART FOUR

Whole-Body Fitness for Pets

· 21 ·

Groom Your Pet for Health

——

Although many dogs and cats clean themselves exceedingly well, most still need our help to keep looking—and feeling—their very best. In fact, regular grooming is an important part of your pet's preventive health program. First, grooming helps keep skin, ears and teeth in top condition. And second, grooming sessions are a perfect opportunity to give your pet a home examination, so you can detect any health problem in its early stages. Also, the touching and massage that go along with grooming relax your pet and strengthen the bond between you.

SPRUCING UP

The best time to accustom your pet to grooming is during puppy- or kittenhood. Groom long-coated varieties daily and short-coated types weekly. Always begin by gently massaging all parts of your pet's body. Talk sweetly and softly to your pet, so the massage is associated with positive attention from you. Use this opportunity to check for any changes in your pet's appearance that could signal trouble. (See chapter 6 for tips on what to look for when examining your pet at home.)

After the examination, you can begin the actual grooming session. Continue talking calmly to your pet as you give it a head-to-toe sprucing up. Pay special attention to the following areas.

Ears. Use a cotton ball moistened with a little alcohol or mineral oil to remove any excess wax.

Teeth. Wipe your pet's teeth with a damp washcloth or gauze pad to remove plaque. You don't have to wait for a grooming session to do this—it's best to "brush" your pet's teeth every day. (For more tips on dental care for your pet, see chapter 10.)

Nails. To be safe, clip only the curved tip of each claw, and always have blood-clotting powder (sold at pet stores) on hand. (More tips on trimming your pet's nails can be found in chapter 22.)

Coat. Brush or comb your pet gently to remove shedding hair and distribute the skin oils throughout the coat. Removing loose hair is especially important in cats to prevent hair balls. Cats can ingest large quantities of hair in the process of licking to groom themselves. Hair balls can accumulate in the stomach or intestines and cause vomiting and loss of appetite.

In long-haired dogs and cats, matted hair can present another problem. Tangles of fur can irritate your pet's skin, and if not removed, they can become so extensive that the entire coat needs to be shaved. To remove a mat, first try to untangle it gently with your fingers, then comb through it. Never tug on a mat—it hurts! If you can't untangle it, use blunt scissors or a mat splitter to cut into the center of the mat. (Be very careful not to cut your pet's skin.)

BATHING YOUR PET

Baths should be given when your pet is dirty or when there is a medical problem, such as parasites or dandruff.

Ordinary baths can be done at home. Medicated and antiparasite baths are best done by your veterinarian or a professional groomer.

Cats rarely need to be bathed because they keep themselves nearly immaculate. Dogs, on the other hand, often need our help. Try not to bathe your dog too frequently, though (no more than once every two weeks). And brush or comb your pet before bathing it. Here are some other tips.

Use shampoos formulated for animals. These are usually gentler than products designed for humans. If you can't find veterinary shampoo, you can use baby shampoo or gentle dishwashing liquid. Never use household cleanser or laundry detergent.

Bathe your pet in a warm, draft-free area. A bathtub is fine if you use a rubber mat to prevent slipping.

Restrain your pet properly. Enlist the help of another person, if necessary, to hold your dog. Cats are most easily restrained during baths by holding the scruff of their necks.

Protect your pet's eyes. Put a drop of mineral oil or eye ointment in each eye.

Use lukewarm water. Wet your pet's coat, then apply a small amount of shampoo. Massage the shampoo throughout the animal's fur, staying away from the eyes and ears. Thoroughly rinse the shampoo from the coat, preferably using a gentle spray attachment.

Dry your pet with a towel. Hair dryers can frighten some pets.

And most important, be patient, and you will both enjoy the benefits.

• 22 •

The Best Way to Trim Your Pet's Nails

—

Scratches on your legs, snags in your sweaters, runs in your stockings and claw marks on your furniture. Do you have a bone to pick with your pet?

If overgrown nails or claws are the problem, a program of regular maintenance is the answer. Keeping your pet's nails fit and trim is easy. And you'll both be happier for it.

Why will your pet be happier? Because your pet's nails or claws do have a purpose (other than to annoy you). Dogs use their hard and resilient nails when they dig to bury things, to make a comfortable depression for themselves before lying down and to burrow under fences or dig up prey. They also use them to paw the ground to leave marks as signals for other animals and for traction when they run.

Cats' long, sharp, retractable claws help them not only catch prey but also climb. And like dogs, cats use their claws to mark their territory.

Like our own fingernails and toenails, pets' nails never stop growing. And few pets have ample opportunity to wear down their nails outdoors, as nature intended. As a result, a pet's overgrown nails can damage your clothing, your furniture or your skin, and they may also be a source of self-injury to your pet.

Long nails can be torn off when they catch on carpet loops, loose floor tiles, window screens and the like, causing considerable pain and bleeding. Occasionally, the nails become so long that they actually curl back into the foot-pads (in a way similar to our own ingrown toenails in terms of pain), resulting in infection. Further, long nails cause the toes to spread, which makes walking difficult. Finally, long nails just don't function well for the special jobs nails are meant to do.

DOWNSIZE THE DAMAGE

Here are some tips on how to keep your pet's nails in check.

Start grooming when the animal is young. Some mature pets don't like having their nails clipped. But if a pet gets used to having its paws handled when young, the problem is minimized. To do that, gently massage your pet's feet every now and then, so the experience holds no anxiety for the animal.

Take it easy. If your pet doesn't like having its nails clipped, try to make the experience pleasurable by combining the procedure with soothing praise and your pet's favorite treats. Move slowly, doing only a few nails at a time. Of course, you need to hold your pet's paws, but do it gently. Try not to squeeze hard or wrestle with the animal.

Buy a good, sharp nail trimmer. Among the many appropriate types is the "guillotine" nail trimmer. It has a metal piece with a hole in it to slip the nail through; then a blade slides across to cut off the nail. I prefer this type because it comes in different sizes and is available with replacement blades, so it stays sharp. (A dull nail trimmer can splinter your pet's nails.)

Other popular nail clippers resemble pliers. They work well, but you may have to look harder to find the right size for your pet. These clippers tend to be larger, and because they're spring-loaded, they may be difficult for people with small hands to operate.

Although small-size nail clippers for dogs work just fine to trim cats' nails, some models are designed especially for cats. I think that the easiest to use on cats is the scissors-type trimmer with a curved blade.

Know where to snip. A nail is a hard, hornlike substance that's shaped like a crescent. It is widest where it attaches to the skin of the toe and narrows to a curved point at its end. The nail has a rich blood supply (the quick) that runs through its middle. Cats and many dogs have white nails through which you can clearly see where the pink quick ends. If the nail is cut too short—to the quick—it hurts the pet and bleeds.

When you trim your pet's nails, always cut well away from the quick. If your pet has black nails you can't see through, make sure that you take off only the thin, curved tip of the nail, away from the quick. Also, trim about every two weeks, so the quick won't grow to the end of the nail.

Snip calmly and quickly. To cut your dog's nails: Hold your dog in the most comfortable position for both of you, with the dog either in your lap or standing on a table. Take hold of each paw gently, and once the dog is holding still, clip each nail quickly.

To clip your cat's claws: Apply gentle pressure to each toe and pad to expose the nail. Then clip quickly.

Don't forget the dewclaws. These grow behind and above the feet and are the equivalent of an animal's thumbs. If your pet has them, it's especially important to trim them because they can't be worn down, since they never touch the ground. If left untrimmed, they curl around and grow into the flesh.

Apply the finishing touches. After you trim your pet's nails, use a nail file to eliminate any rough edges. Some owners use electric nail groomers for this purpose.

Have a product at hand to stop any bleeding. Your pet will let you know if it's hurt during nail clipping, but don't panic. Just stop the bleeding quickly. I use the powdered form of the styptic known as ferric subsulfate (available at pet stores). Don't use a styptic pencil (shaving pencil); the way it must be applied will cause your pet more pain.

CLAW SAVERS

Is your cat the type that could be nicknamed The Shredder? If your furniture is the main target of your cat's claws, don't despair. You have some effective and painless options.

To start, get your cat a scratching post made from rough rope or wood. Place the post near your cat's favorite scratching spot and rub some catnip on the post to attract your cat to it initially.

At the same time you introduce your cat to the post, apply double-sided cellophane tape (cats seem to hate the stuff!) to the places on your furniture that you want your cat to avoid.

Praise your cat for scratching the post. If your cat scratches the furniture, say "No!" Then take it to the post and praise it for scratching there.

After your cat uses the scratching post regularly for a few months, remove the tape from the furniture a little at a time.

If this approach fails, consider "capping" your cat's claws with vinyl nail tips, called Soft Paws, that are applied painlessly with special adhesive. Your vet may have them or, if not, can tell you where to write for them.

• 23 •

Workouts for Pampered Pets

——

The fitness craze has affected not only people but their furry four-footed companions as well. A regular exercise program can be as beneficial, both physically and psychologically, for your pet as it is for you.

Unfortunately, like some people, many of our dogs and cats have grown pudgy from too many treats and inadequate exercise. If you cannot easily feel your pet's ribs when you run your hand along the animal's sides, your pet is overweight. According to the American Animal Hospital Association, obesity affects as many as 40 percent of all dogs and 10 percent of all cats in the United States.

Obese animals experience many of the same health problems that overweight people do. They may be more susceptible to lameness, diabetes, heart and lung diseases and intestinal disorders. Although some animals become overweight due to hormonal imbalance, most cases of obesity are caused simply by ingesting too many calories (especially high-calorie table scraps) relative to the number of calories burned—eating too much and exercising too little. The treatment? Exercise your pet more and feed it a healthy, balanced diet.

EXERCISE FOR BEHAVIOR PROBLEMS

In addition to helping control your pet's weight, exercise helps prevent many behavior problems associated with the boredom of confinement. Many pets destroy items such as books, pillows and furniture in an attempt to entertain themselves. Some pets even injure themselves by excessive licking or biting out of boredom. Most of the so-called hyperactive pets I see are actually underexercised pets. Many of those animals become quite aggressive in their overly exuberant play. Instituting an exercise program is usually very helpful in such cases.

Just as you would see a doctor before beginning any exercise program, have your pet examined by the veterinarian. Your vet can check your pet for any health problems that may be aggravated by exercise (such as heart and joint ailments) and also give you suggestions for your particular pet's exercise regimen.

MATCH THE PROGRAM TO THE PET

Consider your pet's species, breed and age when you design the program. We all know that few cats take to jogging, but it is also true that not too many terriers enjoy swimming. And although you may like to run, walking may be a much safer exercise for your 12-week-old Labrador puppy or your 12-year-old arthritic poodle.

No matter which type of exercise you choose, it should be done at least five days per week. Every day is best.

Very few pets exercise adequately by themselves. You cannot expect your dog to get a good workout by letting it out the back door. And it's not enough to merely provide toys for your cat. Exercise is best done together.

EXERCISING YOUR DOG

Walking. Probably the safest form of exercise for both you and your dog, walking provides aerobic conditioning and muscle toning. Start with brisk walks for 15 minutes each day and gradually increase your time. Stop before either of you becomes winded or sore. Walking is also the best way to begin a safe jogging program, if that is your aim. Always use a leash!

Jogging. If you choose to begin jogging or running with your dog, do so with caution. Always have your dog checked by your veterinarian first. Running with a bad heart or damaged joints could be disastrous. Never run a dog that is under six months old; wait until your pup's muscles, bones and ligaments mature, so they can handle the stress. Try to run on a soft surface to protect your dog's footpads. Don't run on very hot days; it can cause heatstroke and death.

Using bicycles, horses or cars to get your dog to run can be very dangerous. Dogs will overtrain and injure themselves if forced to run that way when they are unfit. Begin slowly by walking. When your dog can handle vigorous walking, begin slow jogs for 15 minutes a day. Increase your time and speed gradually.

Play. Dogs are inherently very playful animals, so this is an easy and enjoyable way for them to exercise. Playing fetch is an excellent way to give your dog a workout without getting yourself too sweaty. Many dogs like to fetch balls, sticks and Frisbees but may need to be taught how to retrieve and give them up. Playing fetch is a good way to develop your dog's coordination, muscle tone and aerobic capacity. It may also help teach your dog to obey you. Chasing moving objects, such as large balls or objects on strings, is another way to get your dog moving.

EXERCISING YOUR CAT

Play. Since cats, like dogs, are inherently playful, most owners can easily use this characteristic in designing a good feline exercise program. Most cats are readily enticed to chase strings and toys attached to strings. Try using a fishing rod–type gadget to pull toys up and down stairs. Never allow your cat to play with string or yarn unsupervised, however. A swallowed string can cause intestinal blockage.

Ping-Pong and aluminum foil balls are great fun for cats. They like to bat them around and chase them. Many cats will even learn to retrieve. You can also make empty paper bags and cardboard boxes available for your cat. Most owners know how much fun their feline friends have with these simple toys. Try to devote at least 15 minutes a day to play.

Outdoor walks. Living entirely indoors extends your cat's life by protecting the animal from cars, dogs and diseases. But it may also lead to boredom and inactivity. Many cats take readily to leash training, especially at an early age, and can then enjoy safe daily walks. I prefer using figure-eight harnesses. Initially, your cat may think it cannot walk with the harness on. First try it on inside without a leash. After your cat adjusts to the harness, attach the leash and take your cat outside. Don't expect your cat to heel like a dog. Basically, let the cat walk you. Aim for at least 15 minutes a day.

Exercise and play not only improve the quality of life for both you and your pet but also help strengthen that special bond between the two of you.

· 24 ·

More on Walking
Your Pet for Fitness

A regular walking program is the safest and most enjoyable way to give you both the exercise you need. But if the longest walk your dog or cat takes is between the bed and the refrigerator, take a close look at your pet. Chances are that it's overweight from eating too much and exercising too little. The cure? A healthy, balanced diet combined with a daily exercise program. Walking is a great way to start.

Daily walks will boost your pet's energy. You may have noticed, especially if your pet is overweight, that it lies around more than it used to and seems uninterested in anything but food. You may have attributed this behavior to age. It could be, however, that your pet is just plain out of shape. A daily walking program will increase the animal's muscle tone, provide aerobic conditioning for the heart and lungs and make it easier for your pet to get around. After a few weeks, you'll probably notice that your pet is more playful, active and happier overall.

We'll focus on dogs here; for information on walking cats, see chapter 23.

STARTING OFF ON THE RIGHT PAW

As with all other forms of exercise, you should have your dog examined by a veterinarian before beginning any walking program. Your vet can check for health problems, such as heart and joint ailments, that could be aggravated by exercise. A vet can also give you specific suggestions for your pet's exercise program. Especially, make sure your dog's vaccinations are up to date, since you are sure to meet other animals on your walks.

WHAT WALKING DOES FOR DOGS

Dogs are naturally very active and playful. In the wild, much of a dog's day is spent hunting for food, defending its territory and playing with other dogs. In contrast, most pet dogs are given all the food they need (usually more) and are confined to the house or yard for most of the day. Generally, they do not have other dogs to play with or a large territory to defend. Dogs need action.

Besides providing exercise for you and your dog, walking is an opportunity for both of you to meet people. Dogs that aren't used to having positive experiences with strangers understandably become afraid of them. And fear often leads to barking and aggression. Daily walks give dogs the important opportunity to meet people in neutral territory and in a positive context. Walks are essential for young puppies (up to six months) to allow them to socialize with all kinds of people. Finally, walking your dog also gives you the chance to meet people. Studies show that dogs act as "social lubricants." Strangers are much more likely to strike up a friendly conversation with you if you're with your dog than if you are alone.

ETIQUETTE FOR DOG WALKING

Always use a leash when walking your dog unless you'll be in an area where there are no cars or other dangers and your dog is well trained off-leash. Your dog doesn't necessarily have to heel perfectly, but it shouldn't tug. Walks are not much fun if you wind up on the ground or with a sore arm afterward.

The best place to learn how to use a leash is in an obedience class, although there are many books that describe heeling techniques. I prefer the technique in which you start the dog in a sitting position, then tell it to walk or heel as you take off. If your dog is a "lunger," the moment it gets to the end of its leash you should reverse direction, quickly jerk on the leash and pat your thigh. When your dog catches up to you, praise it by saying "Good walk" or "Good heel." The change of direction usually has to be done several times, but most dogs get the idea quite easily.

Your leash should be six feet long and made from leather, cotton or nylon webbing. Chain leashes are difficult to maneuver properly. Some dogs train well with just a plain collar, although many need the extra control provided by a choke collar. Incidentally, choke collars are difficult to use properly. (Enroll at an obedience school to learn the technique.) They are made not to choke but rather to correct unwanted behavior.

If you feel you need help getting your dog under control, look for a good obedience instructor or dog trainer. Head collars, such as Gentle Leader or the Halti training harness, are recent developments that make walking your dog even easier. These fit around your dog's muzzle like a horse halter but do not keep the animal's mouth closed.

Try to walk your dog every day, if possible. Dogs are creatures of habit and love routine, so they really miss a skipped walk. Start with 15 minutes once a day and gradually increase your time and pace, stopping before either of you becomes winded or sore. Two 15-minute walks every day is ideal, but one 30-minute walk is also very good.

Of course, nowadays it is unthinkable to walk a dog without the necessary materials for cleaning up after it. Not only is this important for appearance, but it is a worthwhile health measure as well.

• 25 •

Wilderness Walking with Your Dog

———

H eading for the hills for a trek with your dog can be a wonderful experience. Hiking together, you can enjoy seeing the beauty of nature as well as watching your trusty companion have fun in much the way its wolf relatives do. To make your hike the best experience for both you and your pet, though, it's wise to attend to some simple planning and precautions. The tips that follow will help you sidestep some common problems.

Visit the vet. To make sure your dog is in good enough physical shape to endure a hike, check with your vet. While you're there, make sure your pup is up to date on its shots. Rabies, distemper and leptospirosis are carried by wild animals you're likely to encounter, such as squirrels and raccoons.

Your dog should also be on pills to prevent heartworms, which are carried by mosquitoes.

Revitalize Rover. If your pooch has been kind of a couch potato for a while, you should add some physical activity to the schedule. Take brisk walks together twice daily, gradually working up to the distance you plan to hike.

Find out if dogs are allowed. Call the local ranger or other appropriate official before you set out. Even in a rural setting, dogs are not automatically welcome. Many books on hiking areas will indicate whether pets are permitted, so you may be able to get the information you need at your local library.

Lean toward a leash. Even places that do allow dogs often require they be kept on a leash. And it's probably best to leash your dog even if it isn't required. Even a well-trained pet might decide to run off after a wild animal in the excitement of the moment and the new surroundings, and it could get hurt or be lost. So pack a nice long leash and make sure that your dog walks comfortably on it. (Retractable leashes give dogs extra freedom yet allow you to reel them in when necessary.) Don't let your dog run free unless you're sure your commands will be obeyed at once.

Tag your dog. Make sure your pet is clearly identified in case you would become separated. To be safe, attach an ID tag to the dog's collar, even if your pet has an identifying tattoo or microchip. If you are vacationing away from home for a while, write temporary contact information on a bit of paper and slip it into a capsule-type tag on your dog's collar.

Buy the right doggie duds. "A bright orange vest with glow strips is especially important if there are hunters in the area," says Susan Webb, an animal control officer in Wellesley, Massachusetts, who also does search-and-rescue missions throughout the New England states accompanied by her dog. "The vest can also help you spot the dog at night."

Take plenty of water. Dogs are susceptible to many of the same waterborne diseases (such as giardiasis) that afflict people. You can sterilize water effectively by boiling or filtering it or by treating it with iodine disinfectants. But on a brief outing, it's probably easiest just to carry a large canteen of fresh water from home. (Don't forget a lightweight plastic bowl for your dog.) Offer your dog a drink every time you take one. Take more than you think you need—hiking is hard work.

Take snacks. Just like you, your dog will be burning many more calories than usual. Dry food and dog biscuits are both lightweight, and they pack the extra energy your friend needs. Don't ever let your dog eat anything found in the woods.

Keep your canine cool. Hyperthermia, or heatstroke, is a particular hazard for dogs because they do not sweat; they rely on panting to regulate their body temperatures. In addition to offering plenty of drinking water, make sure your dog has ample opportunity to cool off in shady places or in cool, shallow water.

Don't weigh down your dog. If you want your pet to carry a doggie backpack, make sure you accustom the pup to it gradually, as part of the shape-up program. A large golden retriever may be able to tote as much as ten pounds and barely notice it, but a dachshund shouldn't carry anything extra because of that breed's delicate back.

Watch out for rushing water. It's best to put your dog on a leash when crossing streams, says Webb, especially if the dog is carrying a backpack. The current may be too strong. Test the water yourself before you send in the dog. Cross streams at the shallowest point and stay away from waterfalls that might sweep your pet over the edge.

Don't forget first aid. Dogs often cut their footpads or their legs on rocks or trails. So make sure your kit contains gauze pads, gauze and adhesive tape to bandage tender paws for the hike back to the car. "Cuts seem to be most common when there are patches of ice on the ground," says Webb. In those conditions, preventing trouble by using special dog booties makes good sense. They're available in most pet supply stores. These booties can also be helpful for dogs whose pads haven't been toughened up by walking on concrete or on rough terrain.

While you're at the store, pick up a vial of styptic powder to stop bleeding in the event your dog breaks or loses a nail. (Clipping nails ahead of time helps prevent this.)

A pair of pliers may come in handy for removing quills should your dog ever tangle with a porcupine. And consider taking along a snakebite kit, available at hiking and camping stores, as well. (In the event of a snakebite, get your dog to the vet as soon as possible. The smaller size of dogs, in comparison with humans, makes these animals more vulnerable to the venom.)

Keep insects from bugging your pet. Flies and mosquitoes can bother your dog as well as you. The common repellents containing diethyltoluamide, or DEET, are generally safe for dogs if used sparingly.

Get ticks off. A good brushing for your pet after the hike should eliminate any ticks before they adhere and have a chance to transfer diseases (such as Lyme disease). Your dog may also pick up some fleas while hiking. A good mild flea bath for your dog when you get home (preferably with a shampoo with pyrethrins), followed by a flea combing, should get rid of any freeloaders. The bath and combing should also get rid of any ticks you might have missed.

Don't risk ruining the joy of an extended romp with your canine companion by skipping these few simple safeguards. Do what you can to ensure a safe, trouble-free experience, so you'll look forward to hiking with your pet many more times from now on!

Keeping Your Pet Cool through the Dog Days

———

The dog days of summer are no easier on our dogs and cats than they are on us. The hot weather that has us heading for the shade and panting for cool drinks does the same to our pets. But pets don't always complain, so it's easy to forget that they're uncomfortable, too. That's when discomfort can escalate to danger.

Dogs are especially susceptible to hyperthermia, or heatstroke, a potentially deadly condition that often occurs on hot, humid days. Although cats can tolerate higher temperatures than dogs, they, too, can get overheated. Here's how to make sure your pets stay cool and healthy.

Help pets keep their pants up. Unlike people, who rely primarily on sweating to cool their bodies, animals rely on panting. (The only sweat glands that dogs and cats have are in their footpads, and these have no temperature-control function.) When an animal pants, it rapidly draws air into its nasal passages and exhales through its mouth. The evaporation of moisture from the wet surfaces of the mouth and nose help lower the animal's body temperature. Obviously, muzzles should never be used on hot days because they make it very difficult for a dog to pant.

But perhaps less obviously, special care should be taken to keep short-nosed breeds, such as bulldogs, pugs and Pekingese, cool. They are less efficient panters because they have smaller surface areas for evaporation. And note that very humid days are particularly bad for pets, as they are for people, because evaporation is hindered.

Keep the water flowing. Of course, a cool drink helps cool down a body. But more important, just as we lose water when we perspire, animals lose a lot of water from their bodies when they pant. To avoid dehydration, make sure you keep out an abundant supply of fresh water for your pet at all times.

Groom their coats. Although cats pant just as dogs do, they also lick themselves in order to keep cool. The water evaporating from their fur helps provide another means of dissipating heat. To keep this method working at peak efficiency, keep your cat's coat well groomed—free of mats, tangles and burrs.

Give them seasonal hairdos. When the temperature rises, the blood vessels in your pet's skin dilate. More blood flows near the surface, so more heat is lost through the skin. To optimize this automatic cooling system, your pet's coat naturally thins out during the summer months. You can help things along by giving your long-coated dog or cat a short summer crop. (It's a myth that a heavy coat protects pets from the heat.)

Find your pet's cool spot. Most cats and dogs are very good at finding cool places when they're hot. Whether it be under a shade tree, on a tile bathroom floor or lying smack-dab in front of the electric fan, your pet will find the right spot. Your job is to make sure that a cool place is always accessible.

If your dog is outside, make sure there's shade available. You can also turn on the hose and give your dog a puddle to lie in. Let your pet seek the coolest place of its choice whenever possible. Indoors, let your pet join you in front of the air conditioner if it desires. And remember: There are no cool areas in a parked car in the sun. Leaving pets in this predicament is probably the number one cause of heatstroke.

Chill out on exercise. You may have noticed that neither dogs nor cats move very much when it's hot. In fact, most cats seem to just sleep through the hottest part of the day. That's because activity raises body temperature, while inactivity helps keep it low. Do not encourage your pet to exercise on hot days. It's another sure way to bring on heatstroke.

Let your pet eat less. Eating also raises body temperature, so eating less is another method pets may use to keep cool. If your veterinarian has given your pet a clean bill of health, don't worry if your dog or cat eats less on steamy days. Your pet's appetite will pick up when the weather becomes cooler.

Keep your pet trim. A short haircut or thin coat is not going to work very well if there's a thick layer of fat underneath the fur. Fat's insulating value keeps your pet warm in the winter but can be very detrimental in the summer heat.

Obese dogs are much more susceptible to becoming overheated and developing heatstroke. Very obese cats can't reach around to lick their bodies, so they lose this form of cooling, too. If your pet is on the pudgy side, check with your veterinarian for advice on how you can begin a good and safe weight control program.

Keep your pet in top form. Of course, general good health ensures that all the body systems necessary to keep your pet cool are operating properly. Regular annual checkups with your veterinarian are a must.

PART • FIVE
The Well-Behaved Pet

· 27 ·

Get to Know
the Animal Side of Your Pet

When we take dogs and cats into our homes, most of us relate to them as family members. We tend to see our pets as little people. However, they are members of different species and only know how to relate to us as if we were dogs or cats. They use the same sets of signals to communicate with us as they use with members of their own species, just as we use human signals to communicate with them. Problems can arise when these signals are misunderstood.

Take the case of Abby, a 16-week-old puppy who urinated every time her owner, Mike, a large man, walked through the front door or looked directly into her eyes. Mike was convinced that Abby was not housetrained or that she was urinating just to spite him. To correct the situation, Mike rubbed Abby's nose in the urine and spanked her. That made Abby urinate even more.

JUST SAYING HELLO

If Mike had understood normal canine behavior and signals, he probably would have solved the problem without ever seeing me. What Mike needed to know was that urination is often a part of the normal canine greeting display.

During greeting, a dog often assumes a submissive posture to let the other individual, dog or human, know that it has no intention of attacking. In the typical submissive display, the dog crouches, tucks its tail under its body, puts its ears back and smiles. Sometimes the dog may roll over on its side to display its belly. Both of these postures are often accompanied by urination.

Mike's own signals were adding to the problem. A direct stare is a threat signal in dog language and often elicits a submissive display. And punishment, by causing the dog to be fearful and even more submissive, only makes things worse.

Submissive urination in puppies is usually temporary. Until it goes away, owners should try to avoid eliciting it. Greeting should be kept low-key. Loud voices, stares and bold gestures should be minimized. Sometimes crouching to greet the puppy helps.

Other people find that initially ignoring the puppy is the only way to handle the problem. Or try throwing a ball to the puppy when entering the house to elicit play rather than greeting. If all else fails, greeting the puppy outside helps by eliminating owner frustration caused by repeated mopping of the kitchen floor.

SEPARATION ANXIETY

Another very common behavior problem, often misunderstood by owners, is separation anxiety. Dogs who suffer from it destroy their owners' homes, cry or urinate and defecate when they are left alone. The most common explanation that owners apply to these behaviors is spite.

Again, being the humans that we are, we try to understand canine behavior in human terms. But dogs, like their wolf ancestors, are pack animals. When they are separated from their packs, they become anxious and attempt to reunite themselves with the other pack members.

Just as we view dogs as family members, dogs see us as pack members. Therefore, when owners of dogs with separation anxiety leave the house, their dogs become distressed. In extreme cases, they may break down doors or even jump through windows.

Once established, separation anxiety is usually treatable through a program involving a combination of behavior modification techniques and drug therapy. If your dog exhibits signs of separation anxiety, you may want to try a modified treatment program yourself.

Since most dogs start to become anxious as the owners are preparing to leave, the first step is to act as though you're preparing to leave many times during the day—but don't leave. After doing this, your dog should no longer associate your predeparture routine with being left alone and therefore should no longer be anxious before you leave.

If this is successful, the next step is to start leaving your dog alone for very short, varying intervals of time (5 minutes, 10 minutes, 5 minutes, 15 minutes and so on) until your dog can be left alone for two hours.

While you are working the program, it is best not to leave your dog alone for periods longer than the interval of time you are up to. If this is not possible, drug therapy is usually necessary. Because this program can be complex, it's best to seek the assistance of your veterinarian or an animal behaviorist.

However, it is always better to prevent behavior problems from occurring in the first place. Just by understanding that it's natural for dogs to want company, you can try to teach your dog to be comfortable when alone. Owners should avoid being with their dogs constantly, especially at the puppy stage. Caging a young puppy for short periods during housetraining will force separation and help achieve your aim. If you're gone for more than eight hours a day, it may help to have two dogs, so they can keep each other company.

LITTER BOX BASICS

Cats that urinate or defecate outside the litter box are certainly incompatible with human lifestyles. Owners find the odor obnoxious and all too often resort to euthanasia. I get numerous referrals from local shelters and am often the cat's last chance. In most cases, the problem that brought the cat to me is easily remedied.

Take Joanne's Theo, a three-year-old neutered domestic shorthair that had inexplicably taken to urinating on the carpet in Joanne's bedroom during the previous year. Joanne's friends advised her that Theo was just a "dirty cat" and could never be retrained.

Upon investigation, however, it turned out that Theo was just exhibiting normal feline behavior. Cats naturally prefer to eliminate in loose, clean material such as dirt, sand or cat litter. Problems arise when the litter becomes objectionable to them in some way. When this happens, the cat may acquire the habit of using an inappropriate location.

Many cats do not like perfumed litter, liners, hooded boxes or dirty boxes. Joanne had been using deodorized litter and a hood. As soon as she removed the hood and started to use plain clay, Theo began to use the litter box and spared the carpet.

Of course, not all cases are that easy to solve, but litter box problems can often be avoided by taking natural feline behavior into account. (For more tips on taming your cat's litter box problems, see chapter 31.)

Understanding normal animal behavior is essential to the development and maintenance of a good bond with your pet. Although dogs and cats are more like humans than are alligators and aardvarks, we must respect the "animal" in our pets and keep reminding ourselves that we are dealing with other species.

· 28 ·

Stopping Your Pet's Annoying Habits

———

D oes your dog chase cars? Does your cat claw your clothes? Does your pet have some other quirky habits that irritate you? While the behavior may be aggravating to you, it's probably natural for a dog or cat. Such problems are easily solved if you arm yourself with a little understanding of normal animal behavior. Measures you can take to deter the most common of the annoying pet habits are listed below.

When your dog "digs" your carpet. The natural environment for dogs and their wolf ancestors is the outdoors, where they instinctively dig out a depression and trample the grass around it to make a comfortable, soft, cool sleeping area. Since there is no dirt or grass in your living room, your dog applies this inherent behavior pattern to the next best thing—your carpet. To stop this behavior, give your dog a soft bed or blanket of its own, show the bed or blanket to your dog, and praise your dog for "digging" around in it. Whenever your dog scratches at the carpet, say "No!" Then call your dog over to its bed and praise the animal for lying there.

When your dog runs after traffic. The habit some dogs have of chasing cars and bikes can be very dangerous for all involved. Scientists aren't sure why dogs do this, but some think that fast-moving conveyances resemble prey running away, and the dog's natural response is to chase them. Another possible explanation is that spinning tires and fast-moving objects simply excite dogs and they love to chase them.

If your dog is a chaser, take it to a good obedience class, in an atmosphere free of distractions, where it will learn to sit on command and stay in place reliably. Then take your dog to a place where there are bicycles or cars, but keep it some distance away from them. Practice your commands there, using a lot of praise and food rewards. (Make sure you have your dog on a leash at all times.) Gradually, get closer and closer to the exciting objects as you practice, until your dog is more eager to listen to you than to chase. If your dog tugs at the leash, respond immediately with a sharp jerk on the leash and a firm "No!"

When your pet tries to run out of the house. Does your cat or dog try to slip out the front door every time it's opened? It's understandable; being outside is much more exciting than being inside. Some pets just want to explore and play, while others may want to defend their territory. Some simply want to go where you're going. For dogs, the solution is easy. Always tell your dog to sit and stay before opening the door. Make sure you praise your dog afterward. And when leaving together, make sure your dog stays and allows you to pass through the doorway first, then call it to follow.

Cats are a little more difficult to control in this regard. A method that often works is to arm yourself with a plant mister filled with water every time you open the door. If your cat starts to dart, give it a spritz. Do this a couple of times, and your cat will know you mean business as soon as you pick up the bottle.

When your cat "kneads" you. The kneading cat is displaying a behavior that it regularly performed as a kitten. In order to stimulate milk release from its mother, a kitten treads with its front paws around the mother cat's nipples. Some grown-up cats are reminded of their mothers' soft, inviting abdomens when they feel soft, warm surfaces. So even as adults, they revert to that nursing mood and begin to knead on *you*—sometimes shredding your favorite outfit in the process. To discourage this behavior (sometimes those claws are painful!), give your cat a soft surface of its own. My cat loves to knead one of my old sweatshirts, but a soft blanket does nicely, too.

When your dog disturbs the peace. They say some dogs just like to bark and do so in response to almost anything: the doorbell, a person passing the house, even a leaf floating by. Some dogs bark aggressively in defense of their territory. Others have learned to get attention or food from their owners by barking. Still others bark only when left alone. These are usually dogs with separation anxiety, and they bark and whine (just as puppies do when separated from their mothers) in an attempt to get reunited with their owners.

If your dog is a nuisance barker, try to figure out why. If your dog barks in response to almost anything, make sure to keep the animal with you, so you can supervise it—don't leave your dog outside unattended. After it barks a few times, let your dog know "That's enough!" and give it a correction at the same time, either with a sharp jerk on the leash or by shaking a "rattle can" (an aluminum can containing about 20 pennies). Then call your dog to you, tell it to sit and stay, and praise it for obeying you.

If your dog barks in order to get your attention or for food, never give it either of those immediately after the animal has barked. Wait until your dog is sitting and staying quietly before you give it your attention or a food treat.

If you think that your dog barks because it is anxious about being left alone, get your veterinarian or a qualified animal behaviorist to set up a behavior modification program that is likely to help. Such a program usually involves a series of departures and quick returns to teach your dog that you will come back and not to be anxious about it when left alone. You can try behavior modification following the steps outlined in chapter 27.

When your cat wets your bed. Cats that urinate on beds present a frustrating and perplexing problem for their owners. Some cats urinate only on one person's bed or in the place where that person sleeps. (This often happens when a new spouse moves into a home.) In most cases, the cat is marking with urine in response to the invasion of its territory by the strange-smelling newcomer. The problem may persist until the cat accepts the new person.

If you have to deal with this problem situation, first see your veterinarian, so she can make sure that your cat doesn't have a medical problem. If your cat is healthy, try making the litter box absolutely wonderful for it. (For more help in solving litter box problems, see chapter 31.)

When your dog nips at heels. Shetland sheepdogs and Border collies, bred to herd sheep and cattle, commonly do this. Such dogs are merely doing the jobs that they were born to do. When there are no sheep around, they herd the next best thing—their owners. Other dogs, especially puppies, just like to chase and nip moving feet because it's fun. Others nip at their owners' feet only when the owners are leaving the house. These dogs usually suffer from separation anxiety.

To solve this problem, try the same basic strategy used to keep dogs from chasing cars. Start by teaching your dog to sit and stay, rewarding it with praise and food. Next have people walk by your dog, first at a distance and at a slow pace. Then gradually reduce the distance between the dog and the person and quicken the pace. Praise your dog for not nipping as the person goes by, but correct it with a jerk on the leash or a rattle can if it tries to nip. Also practice walking to the door while your dog is in the sit-stay attitude.

Never run from your dog if it starts to nip at your heels; this is exactly what the animal wants, and your action only reinforces the behavior. A better response is to stand still and give your dog the sit-stay command.

When your pet devours your plants. It seems odd that pets normally thought to be carnivores would want to eat vegetable matter. But contrary to popular belief, grass and other plants are part of the regular diets of free-roaming dogs and cats. Some scientists feel that plants serve as a type of bulk laxative, helping the undigestible parts of prey animals (bones and hair, for example) pass through the intestinal tract. Dogs also often eat grass when they're nauseated, to initiate vomiting. Of course, some dogs, especially young ones, just pull plants apart in play.

To discourage a plant eater, try one or more of the following.

• Keep all plants out of your pet's reach, if possible.

• Booby-trap plants by using black pepper, Tabasco or other safe, bad-tasting substances that can be found at your pet store.

• Include some vegetables in your pet's diet (introduce them gradually to avoid digestive upsets). Try green beans, carrots and lettuce first.

• Give your cat its own plants. Grow a pot of "kitty grass," readily available in pet stores.

You will find that most pet behaviors that are obnoxious to humans have a foundation in the pet's inherent nature. With patience and a little know-how, you can discover the source of the problem and devise a way to end it, so you and your pet can go on to enjoy each other.

• 29 •

Reclaiming Your Comfort Zones from Your Pet

▬

Does your cat curl up on your comforter? Does your dog doze on the divan? If so, you're not alone. But you can reclaim your bed or furniture by using an approach that takes advantage of your pet's natural inclinations. Here's how to choose the right bed for your pet and how to get your pet to use it.

DOG HEAVEN

First do everything you can to the bed seem absolutely wonderful to your pet. Just like us, animals want to feel comfortable and secure when they sleep. In the winter, they seek warm places; in the summer, they prefer cool spots. When dogs and cats sleep outside, they usually pick soft, quiet, protected areas. Dogs sometimes dig depressions in soft dirt where they bed down. Cats frequently choose hidden locations.

Most pets like to sleep close to their owners. That's especially true for dogs, the natural pack animals. When they live in human households, the family is their pack. So if you want your pet to enjoy its own bed, make sure you put one in each of the rooms where the family spends most of its time (the bedroom, kitchen and living room, for example).

The bed should be in a quiet area, out of the traffic pattern. A corner is ideal, since it allows your pet to feel protected on two sides.

What is the best type of bed for your pet? That depends on whether it is a dog or a cat, its age, its health, maybe even the environmental temperature. For example, many dogs like the "big pillow" beds because they're soft and because dogs can easily hollow out depressions in them. The beds are available with a synthetic, downlike filler, which is supposed to retain body heat; cedar chips; or a combination of both.

Red cedar chips have been touted as a natural repellent for fleas and ticks, but there is no proof to back these claims. Most of the big pillow beds come with washable covers and in a variety of sizes and colors.

Other popular beds, especially favored by smaller breeds of dogs, are those constructed of foam covered with soft fabric, fake fur or artificial sheepskin (manufacturers claim that sheepskin keeps pets warm in winter and cool in summer). These beds have high sides to make your dog feel protected. They may not be ideal for puppies, however, for the foam is easily destroyed by sharp baby teeth at play. The covers are removable for laundering.

CAT'S CRADLE

The ideal cat beds take advantage of the feline inclination toward privacy as well as safety. Look for those that have high sides or are shaped like tunnels or tents. They should be plush and soft because many cats like to knead before they lie down to rest and the soft surface indulges this behavior.

For older pets and those with aching joints or other orthopedic problems, there are beds that may help ease the pain or stiffness and help the animal sleep better. One type is electrically heated and operates much like a heating pad. Or if you prefer, a special pet heating pad can be purchased and placed under any pet bed to warm it. (Do not use heating pads intended for people; they get too hot for pets, and they're not designed for pet safety.)

Another type of pad was originally designed to reduce pressure on joints in both bedridden humans and animals. It's basically an egg-crate foam pad covered with a soft, absorbent material, such as artificial sheepskin. Common in many veterinary hospitals, it's the bedding most often used in the intensive care unit at Angell Memorial Animal Hospital in Boston. Alicia Faggella, D.V.M., director of the unit, feels that these beds are also beneficial at home for older pets and pets with arthritis. Although they don't slow the progression of arthritis, they do keep the animals warm and their bones well cushioned. Dr. Faggella noticed that her own arthritic dog was less stiff and seemed more comfortable after he started to sleep on one of the egg-crate beds.

OUT ON THEIR OWN

Of course, just buying your pet a wonderful new bed doesn't guarantee that your dog or cat will use it. Some pets must be trained to indulge themselves in their latest luxury. With a new pet, the process is relatively easy. Never allow the animal to be with you in bed or on the furniture. And of course, praise it whenever it goes to its own bed. A food treat given when the dog or cat is already in its bed is also helpful. Once the pet is attached to its bed, you may occasionally allow it to join you on the furniture or bed, but only when invited.

The training process is a little more troublesome for the pet that has enjoyed sleeping in beds and on furniture for years and is no longer allowed to do so (because of a new couch, new spouse or new allergy, perhaps). As people do for their new pets, you must provide the most perfect bed that you can, then reward your pet with praise and a food treat for using it. And never allow the animal on your bed or the furniture. If your dog does venture onto the forbidden areas, respond immediately with an emphatic "Off!" and a quick but gentle push onto the floor. Follow with "Go to your bed" (which should be nearby) and praise the dog when it does. Most cats respond well to "Off!" accompanied by a spray of water from a plant mister whenever they venture onto forbidden territory.

When you're not there, or when you are sleeping, your pet may try to sneak up to its old resting place. For those times, think of ways to make the beds and furniture inaccessible or unpleasant. Try closing doors or putting up pet gates. Other safe, aversive booby traps for sneaky pets include double-sided cellophane tape, aluminum foil, "rattle cans" (aluminum cans filled with coins) and commercial pet repellents.

Both you and your pet are entitled to rest in peace and comfort. With the right equipment, plus some ingenuity and determination, you can have your desire practically overnight.

· 30 ·

Taming
the Home-Alone Animal

———

Oliver is a friendly, playful two-year-old Samoyed owned by Laura and Ron. On weekdays, Laura and Ron go to work, leaving Oliver at home alone for eight or nine hours. Most of the time Oliver seems fine, but a couple of times a week he finds something, such as a pillow or knickknack, to chew on. Laura wonders if Oliver is being spiteful or maybe is just plain bored. In frustration, she asks what can be done.

Laura's complaints are very common among working pet owners. Many feel that their pets are "getting back at them" for being left alone. Unlike their human companions, however, dogs and cats are not spiteful.

Dogs are social animals like their wolf ancestors, who live in packs. Dogs are genetically programmed to live with others, so they tend to have problems when left alone for long periods. Cats, on the other hand, are naturally more solitary animals and generally do well when alone. Some cats, however, seem to be much happier when they have company.

133

THE PLAYTIME REQUIREMENT

Many behavior problems of some pets stem from their inherent need for play. This is especially true for younger animals. Left to their own devices, many animals will design their own games, much to the dismay of their owners!

Even though you may not be able to spend a lot of time with your pet, you can offer it quality playtime. Try to romp with your pet 15 to 30 minutes every day. Most pets that are played with regularly are happier, thinner and younger acting. If you have ever watched dogs and cats wrestling with members of their own species, you know that they do not play gently. Therefore, play should be relatively energetic.

Exercise is another good way to offer quality time to your pet.

WHEN YOU'RE AWAY FOR A WHILE

If you're gone for more than eight hours a day, you might consider hiring a dog walker to provide extra stimulation and exercise for your dog. Many times schoolchildren who don't have their own dogs are willing to help out. Professional dog walkers are also available in most urban and suburban areas. Check with your veterinarian for some references.

Although most dogs and cats may spend much of their time alone sleeping, they still enjoy a few toys to keep them entertained. The presence of toys alone should never replace active play, however.

If your pet is destroying your house in play, pet-proofing may save you quite a few bills as well as protect your pet. This is especially important for puppies and kittens. Keep your pet confined to a safe area—the bathroom, the kitchen or a cage. Make sure all food, valuables and dangerous objects (such as electrical cords) are out of reach. For puppies who love to chew on table legs and cabinets, bad-tasting substances such as Bitter Apple (available at most pet stores) sprayed on those objects twice daily often help deter eager teeth.

If your pet engages in undesirable play or chewing in your presence, reprimand it with a firm, loud "No!" accompanied by a sharp hand clap. Squirting water at your cat with a plant mister often works well to halt unwanted behaviors. Then help your pet by directing it to the nearest acceptable toy and by praising the animal with a gentle, pleasant voice tone.

IT'S NICE TO BE MISSED, BUT . . .

Some pets become extremely anxious when left by themselves. They don't play when left alone but do things out of anxiety. Separation problems are much more common in dogs than in cats because dogs are much more social. Dogs with separation anxiety may bark nonstop, urinate, defecate or destroy property. They never behave this way when you're in the house, only when they're alone. They typically chew and scratch doors and windows. Many dogs also whimper and look depressed while their owners are preparing to leave. Cats with separation anxiety may spray, stop using their litter boxes or become destructive.

Problems due to separation anxiety typically occur soon after owners leave. And unlike play destruction, which occurs irregularly, problem behaviors due to separation anxiety usually occur daily. (To help your dog overcome separation anxiety, you might want to try the behavior modification program outlined in chapter 27.)

Punishment for your pet's destructive acts that occur while you're away is both ineffective and cruel. To work, punishment must immediately follow the unwanted behavior.

The best time for prevention to begin is before you acquire a pet. First consider your lifestyle. Do you have the time to exercise a dog adequately, or would a cat be better for you? How many hours would your pet have to spend alone? Studies have shown a tendency toward separation anxiety in dogs left alone for more than eight hours. So if you are gone longer than that, consider a cat.

Separation anxiety is less common in multidog households, but don't rush out and get another dog in order treat a separation problem. It is *you* to whom your dog is attached. Besides, when two dogs start to play, pillows and shoes make great tug-of-war toys!

Cats and dogs are usually not ideal company for one another. While most cats enjoy living with other cats, some will not tolerate other cats in their territory.

Finally, to prevent separation anxiety, try not to spend long periods with your pet without sometimes leaving it alone. For example, if you're unemployed for a lengthy time or work in the home, try to get out of the house and leave your pet alone for short periods.

· 31 ·

Solving Your Cat's Litter Box Problems

▬

Litter box mishaps are the most common behavioral complaint owners have about their cats. And who can blame them for complaining?

If your cat has a litter box problem, though, try to remember that it is not being spiteful. In most cases, there is a clear-cut reason why your cat has changed its habits, and the problem can be corrected.

First order of business: Determine the cause. That means a trip to the vet, so he can rule out problems such as cystitis, diabetes and diarrhea. If your doctor finds a medical disorder, he will prescribe a treatment that should cure the problem. If there's no medical cause, your cat's problem is behavioral.

MARKING TERRITORY

When marking, or spraying, cats are leaving calling cards. The urine they deposit within their territory lets other cats know who they are, where they are, their sex and if they are ready for mating. This behavior is most common in nonneutered male cats and in female cats in estrus (the fertile period). A neutered cat may still spray, however, if it is bombarded with stimuli that elicit territorial competition and sexual behavior. The sights, smells and sounds of outdoor cats can elicit marking in an indoor cat.

Here are some treatments that may be effective.

• Consider neutering. This is likely to work even in fullgrown cats.

• Make marking sites unpleasant to cats. Place lemonscented air fresheners or aluminum foil at trouble spots.

• Prevent access to outdoor cats.

• Weigh the value of drug therapy. The most effective treatment for marking includes drug therapy in combination with the above. Female hormones and tranquilizers are most commonly used for this, but your pet must be monitored for side effects while taking them.

INAPPROPRIATE ELIMINATION

Most cats that forsake the litter box do so because they've developed a dislike for it and a preference for another location. Common causes are dirty litter, a new or perfumed litter, a hood on the box or being punished or medicated near the box.

Cats' choices for new toilet areas can be disturbing. The corner of the living room rug and the bathtub are common selections.

The basic premise of treatment is to make the litter box attractive for the cat, while making the inappropriate area unpleasant. To make boxes more desirable:

• Use plain clay litter. Avoid deodorized or perfumed types. If your cat does not seem to like clay, try newspaper or wood shavings.

• Put two to three inches of litter in each box your cat uses.

• Using a litter scoop or spoon, remove both the urine and feces at least once daily. Always follow this step by adding fresh litter.

• Change the box completely if you can detect odor when you put your nose close to the freshly scooped box. (If the smell puts *your* nose out of joint, consider your cat's sensitive snoot.)

• When the box is changed, wash it out with a mild detergent, such as dishwashing liquid. Avoid harsh household disinfectants. Some can be toxic to your pet.

• Try to provide one litter box per cat. The boxes will stay cleaner.

• If the box has a hood, remove it.

• Move the feeding area away from the box.

• Move the box to a different location. A more accessible or quieter area may help. Or place the box in the area the cat is already using.

At the same time, do what you can to make the inappropriate toilet area intolerable for your cat. Here are some simple but effective tactics to try.

• Put a lemon-scented air freshener in the area.

• Feed or play with the cat there.

• If your cat targets a carpeted area, cover that place with a thick plastic drop cloth.

• Leave a few inches of water in the bathtub, if that's the scene of the crime.

Also try rewarding your cat with praise and a special treat every time it uses the litter box, while discouraging it from the inappropriate area with a firm "No!" and a clap of your hands. Never hit your cat or punish it after it has left the evidence, however.

Once your cat is using the box again, leave everything in place for a month. Then you can begin to change things back gradually. Most households are back to normal within three months.

• 32 •

Choosing the Best Litter Box Products

—

When it comes to controlling litter box odor, some cat owners go to extremes. They tuck the cat box in the deepest, darkest recess of the basement, with a large plastic hood over it and a pine-scented air freshener plastered on the side. Then they're surprised when Kitty soils the new white rug. If your cat doesn't like the arrangements you've made for its toilet habits, it will let you know soon enough.

One reason cats have achieved the exalted status of America's most popular pet is that they're easy to take care of. Unlike dogs, who must do their business outside, most cats readily use litter boxes. Cats instinctively bury their urine and feces, which makes a loose material like litter their preferred toilet.

Many owners object to the appearance and odor of used litter. In response, the companies that produce this product have developed a variety of methods to neutralize, deodorize and otherwise eliminate litter box smells. Unfortunately, some manufacturers put so much emphasis on pleasing owners' noses that they forget the litter has to agree with cats, too.

Of course, nothing controls odors better than a clean litter box. It makes sense: Remove the source of the odor, and the smell is gone. Some suggestions to help keep the litter box odor-free for you yet appealing for your cat can be found in chapter 31.

THE BEST TYPE OF BOX TO BUY

My experience tells me that cats prefer plain plastic litter boxes without hoods. When I see a cat that is showing signs of litter box aversion (it chooses the bedroom carpet over its box), I often experiment to find which kind of box the cat prefers. Of the more than 1,000 cats I have treated for this problem, only 2 preferred litter boxes with hoods.

While that doesn't bode well for boxes with hoods, you might feel so strongly about cat box odor that you're willing to give them a try.

Ultimately, the problem with hoods is that owners simply may not see or smell soiled litter boxes, which means they might be less likely to clean them. Consequently, their cats are also less likely to use the boxes.

If your concern is keeping litter from being scattered outside the box, try a box that comes with a wide lip. But if odor is what bothers you, a hooded box may be counterproductive unless you're very careful about keeping it clean.

WITH CATS, TEXTURE COUNTS

All cat litter ads claim their products control odor. No available studies support one manufacturer's claim over another. But one study suggests that it might be wise to take your cue from your cat.

New York City animal behaviorist Peter L. Borchelt, Ph.D., examined cats' preferences for different types of litter materials. He compared several brands of popular clay litter, the sandlike clumping litter and plain sandbox sand. Left to their own devices, the cats most often chose the sandlike litter.

The sandlike litter is a relative newcomer to the litter box game. Made from finely ground clay, this litter is soft on cats' paws, and that may be why cats prefer it. This particular litter is also easy for owners to clean. When a cat urinates in it, the urine clumps into a small patty that can readily be removed with a litter scoop. With many, but not all, of these products, the urine clumps can be flushed down the toilet. (Product labels tell you if a brand is flushable. Do not flush any kind of litter if you have a septic tank.) If both urine and feces are scooped daily and fresh litter is added, the box hardly smells.

Many litters, including the scoopable types, come with deodorizers, such as baking soda or even baby powder. This may be more for the owner's benefit than for the cat's. If litter boxes are properly cleaned, deodorizers shouldn't be necessary. Litter choice experiments among my feline patients show that most prefer their litter unscented. However, some cats that are started on a certain deodorized litter as kittens prefer it when they're full-grown cats.

But what does it matter which litter you buy if your cat refuses to use it? You may need to experiment to determine which litter your cat prefers. Just put two boxes side by side. In one box put the litter you're now using, and place the test litter in the other. Try all types—deodorized and undeodorized, plain clay and sandlike clay.

ADDITIVES AND LINERS THAT MAY DO THE TRICK

If odor is still a problem after all your attempts to curb cat box aroma, you might try one of the numerous cat litter additives. Some contain live bacteria and enzymes that break down the odors. Others have very efficient absorbents, such as an ammonia-absorbing resin. Again, there are no studies vouching for the effectiveness of these products. Some have strong odors, which might be unpleasant for some cats.

Before you decide to use one of these products regularly, see if your cat minds the new smell. The signs of displeasure will probably be subtle. For example, when I tried a deodorant in my cat's box, he stopped covering his feces. As with different types of litters, you might experiment to see whether your cat tolerates one additive or another.

Finally, some people find plastic cat box liners helpful. Liners make cleaning litter boxes easier, but some cats do not seem to like them. One popular liner is fashioned after disposable diapers, soaking up urine with an absorbent pad. The manufacturer recommends changing the liner every two weeks. One negative: These liners make scooping clumpable litter more difficult. Make sure that your cat approves before relying on them exclusively.

In the end, you may have to fall back on that most effective way to curb the unpleasant odor of cat litter: Keep the box accessible and clean.

How to Tame a Jealous Pet

——

Adam, a three-year-old cat, had impeccable litter box habits—until his owners brought home their new infant son. Adam began "extending" his litter box until it included the entire living room.

Francis, a beautiful collie, never seems to get enough attention. When his owners play with their daughter, Francis pushes between them—and because of his size, he really gets in the way.

Bridget, a long-haired cat, had the apartment to herself—until her owner decided to get a kitten. Bridget chases the kitten every time it appears. Now the kitten is so terrified that it lives under the bed.

Are these animals jealous? This kind of behavior does indeed resemble the way human beings act when they're jealous. But once you understand the world of dogs and cats, you'll see that what they're feeling is a little different. If Adam were a wild cat and a strange new creature with unusual smells and sounds invaded his territory, he would probably mark his area to warn it away. Cats mark with urine, of course. Unfortunately, in your home, your cat's territory may include an expensive carpet.

Bridget's reaction to the new kitten can also be explained by territoriality, not jealousy. She doesn't want the kitten on her property, so she bullies it into hiding.

What about Francis? Dogs are very social animals: They play, hunt and eat in packs. Francis is behaving like a normal dog. He wants to participate in the family fun!

YOUR PROBLEM CAN BE SOLVED!

Even though our pets aren't feeling true jealousy, living with more than one animal, or introducing new individuals into the household, can raise similar problems. Luckily, there's a lot a pet owner can do to promote harmony and more acceptable pet behavior.

Within packs, dogs set up a dominance hierarchy—a pecking order. In a family, a dog sees both humans and other animals as members of its pack. The best place for a human to be is on top. Otherwise, the dog can get away with disruptive behavior.

How do you make yourself the leader of the pack? Usually, it happens naturally, but if not, obedience training can help. Find a class based on reward—not punishment— and have all the family members work with the dog. Make your dog sit before it receives attention of any kind: a stroke, a treat, a meal, a walk, opening the door, throwing a ball. The message: Nothing in life is free. This lesson can help keep order in your family life.

Dogs don't have democratic instincts. They're happiest in a stable hierarchy. So if there are several dogs in your home, you should support the order they establish. Give the dominant dog food first, attention first and first right-of-way through tight areas such as doorways—these status symbols are important in dogdom. The subordinate dog understands its place in the hierarchy and won't feel cheated or jealous. If you want to bring in another dog, remember that dogs of the same sex have more conflicts, so an animal of the opposite sex is a safer bet. Also, neutered dogs (and cats) fight less.

If possible, introduce the two dogs somewhere away from home—such as a park. Once they become friendly on neutral territory, bring the new one home.

DEFEND YOUR CAT'S LAND

Unlike dogs, cats are loners by nature. They hunt alone, and except for a mother and her kittens, they live alone. And as we mentioned before, they defend their territory, whether it's in the wild or in your home.

The first thing you can do to prevent problems is to limit your pet population. Overcrowding leads to marking and aggression. If you already have one cat and want another, a playful kitten may not be the best choice. Older cats are often annoyed by kittens but may accept other adult cats easily.

The introduction of a new cat should be done gradually. Keep the two cats separated for a few days. Rotate them from room to room, so they become familiar with each other's odors. Then allow them to be together. Probably there will be a few days of hissing and growling, but it should cool down.

If you have several cats, offer many choices of the things they tend to fight over: eating places, beds, litter boxes and scratching posts. Give your cats several places to hide: "kitty condos," paper bags and boxes. Don't force your cats to share.

If your cats or dogs start to fight, never try to break it up with your hands—it's the surest way to get bitten. Instead, pour water on them or maneuver a pillow or blanket or another object between them.

NEW PEOPLE, NEW PROBLEMS

If the new family member is an adult (such as a spouse), make sure that all the initial experiences are positive. The new person should speak gently to your pet, play with it and feed it treats at the first meetings. If your pet is afraid of the new person, don't allow that person to hit or punish it in any way. That only makes matters worse.

There are many things you can do to ease the introduction of a new baby. One of the best tricks: Expose your pet to the baby's sounds and smells first. Bring home diapers or blankets from the hospital, as well as tape recordings of the infant's cries, a few days in advance. When mother and child come home, the pet should be greeted without the child present. Then gradually bring them together. (For more tips on helping your children get along with the family pets, see chapter 34.)

TRY TO RELAX!

Remember that change of any kind is stressful for an animal—cat, dog or otherwise. So whether the new family member is a person or an animal, that's enough stress—it's important to keep the rest of your pet's life as close to the old routine as possible. If you fed your dog twice a day and walked it three times a day, followed by a special treat, continue to do so. If your cat always slept on your bed, don't kick it out now.

And try not to be overly anxious. Animals can sense our apprehension, which can make a difficult situation even tenser. In most cases, the animals seem to work things out by themselves despite the situation.

• 34 •

Helping Pets
and Children Get Along

——

Amber was a six-year-old golden retriever owned by Mr. and Mrs. Grant. When they called me, their son, Jason, was 18 months old, and their daughter, Jennifer, was just 1 month old. Amber had always been especially good with children, but soon before the Grants consulted me, she had started growling at Jason. Amber never growled at Jennifer, nor had she growled at Jason when he was an infant. The Grants called because they were concerned that Amber might hurt Jason or even "turn" on Jennifer. They wanted to know how to help Jason and Amber become better friends.

Such a situation is not uncommon in families with small children. Although care should be taken to introduce your dog to a new baby slowly and under control, most dogs readily accept infants. If problems arise, it usually happens when the child begins to crawl and walk.

There are several reasons why a dog may start to growl at, snap at or even bite a small child in the family. (Aggression toward children outside of the family will be discussed later.) In Amber's case, the signs of aggression were a result of fear. Jason, like most boys his age, was extremely active and often unaware of his own strength. When Jason began to crawl, he enjoyed chasing Amber and pulling on her tail. When he started to stand and walk, he often used Amber for support when she was nearby. A few times Mrs. Grant heard Amber yelp. After a while, Amber began to avoid Jason. But when Jason started to run, Amber could not get out of his way fast enough, so she began to growl at him.

Many parents believe that dogs should tolerate just about anything from young children. But dogs feel pain just as much as we do. Careful supervision of young children and pets, to protect both the child and the pet, cannot be overemphasized.

PEACEFUL COEXISTENCE

To prevent a fear-induced aggression problem, start teaching your child early to pet gently with an open hand. Try to make him understand that the dog is not a toy but a living thing that feels.

I always recommend that owners give their dog a place of its own, either in the corner of a room or in a crate. This place should be off-limits to the child when the dog is resting or sleeping there. It is best to teach a dog to go to its place on command at a young age, before the children are born.

Another reason that dogs may show aggression toward toddlers and young children is related to dominance. Dogs are naturally social animals that live in packs. Within a dog pack, a stable pecking order, or dominance hierarchy, sets up. This hierarchy is maintained by threat signals, such as growling, snarling and snapping. Overt fighting, where the dogs get hurt, rarely occurs.

When we take dogs into our families, they naturally relate to us as pack members and take their places in a human/dog dominance hierarchy. In most families, the people are clearly dominant over the dog, and there are no problems. In some families with young children, however, the dog views the child as a subordinate and relates to the child as another puppy. In these cases, the dog may growl if the child approaches the dog's food dish, disturbs the dog while sleeping or resting or attempts to climb on a parent's lap. Sometimes the dog will growl at the child if hugged too hard or stared at eye to eye. In most cases, as the child grows up, the dog begins to view the child as dominant, and the growling diminishes. But until then, it is very important to supervise the child/dog interactions. *If you cannot supervise, separate.* Child or pet gates are wonderful inventions. It can't be repeated often enough: Teach your child not to go near the dog while the dog is eating, resting or sleeping.

Unfortunately, some dogs attempt to assume dominant positions over several or, in some cases, all family members. These dogs may be very dangerous to keep with children.

In *any* case, aggression toward young children in the family is potentially a very serious problem. If you have such a situation, seek advice from your veterinarian or a qualified animal behaviorist.

WHEN PLAY GETS TOO ROUGH

One of the most common complaints I receive about pets and children concerns playful aggression. Young dogs often jump on children, happily biting and chewing on their arms and ankles and chasing them around the house and yard. This play aggression is often accompanied by growling and barking, but the sounds are different from those of serious aggression (such as the ones your dog might make when the mail carrier comes to the door). Dogs at play often assume "bowing" postures and rarely break the skin.

When a dog plays too rough, my best advice is to teach your children to ignore the animal. They should fold their arms, for example, close their eyes and remain perfectly still. While your dog responds well to rewards for forms of acceptable play such as fetch and Frisbee, punishing rough play often backfires. Any attempt to push the dog away, knee it or even hit it is often perceived by the dog as part of the game. That may perpetuate the very behavior you want to discourage.

Young cats often play aggressively, too. They stalk, chase, pounce upon and bite moving objects, even if those objects are attached to human beings (hands and feet, for instance). In dealing with such problems, I often recommend getting the cat a feline friend to play with. If that's not feasible, then I suggest providing the cat with fun toys, such as Ping-Pong balls and string toys. If the problems persist, punishing inappropriate play either by spraying with a plant mister or by making loud noise often helps.

Even if you don't have children at home, it is important to be able to trust your pet around children. Grandchildren, nieces and nephews, friends' children and the newspaper carrier should feel comfortable. When your pet is young, you should try to introduce it to children of all ages. They move differently from adults; they are louder and sometimes seem threatening to a pet. Have the children gently pet your dog or cat and feed it delicious treats. If your pet is older, you can still expose it to children, but do so more gradually.

Experts agree that most of the animal bites children receive each year, whether from their own pets or other animals, can be prevented. Instilling in children a sense of respect for animals as animals, rather than as toys or little people, is of major importance.

· 35 ·

Help for Noise-Sensitive Dogs

—

At the first boom of thunder or firecrackers, Sarah, a six-year-old pointer/Labrador mix, would tremble and pant. Sometimes she'd climb into the bathtub and "dig." Or she'd claw at doors, windows and floors, trying desperately to get away from the loud noise.

Sarah is typical of the many dogs that have what animal behaviorists call loud-noise phobia. Phobias are abnormally severe fear responses. And while scientists aren't certain what causes phobias in animals, they suspect there may be a genetic link. Loud-noise phobias, for example, are most common in the hunting breeds, such as Labradors and Irish setters.

UNLEARNING FEAR

Unfortunately, puppies who display phobic behavior aren't likely to outgrow the problem. But many dogs are treated successfully with a technique called behavior modification. The process involves exposing your dog, gradually, to loud noises. (Hunters use this method on gun-shy dogs.)

To try the technique, you will need a tape recording of some fear-producing sound. Thunderstorm recordings are usually available at large record stores. Or you can record Fourth of July fireworks or a backfiring motorcycle yourself.

To begin, put your dog in a relaxed state. Choose a place where your dog feels safe, such as its bed or a special rug. Then practice some commands, such as "sit," "down" and "stay," and reward your dog with its favorite food treats. Conduct these training sessions for about 20 minutes a day for a week. By the end of that time, your dog should associate these sessions with pleasure.

Now you're ready to introduce the tape recording during a training session. Play the tape so softly that you can barely hear it. (Your dog's ears are much more sensitive than yours.) Continue to practice your commands while rewarding your dog every five or ten seconds. After at least five minutes—and when it is apparent that your dog is not paying any attention to the recording—turn up the volume a bit and continue to practice your commands.

If your dog shows any signs of fear, such as panting or trembling, turn the volume back down. (Do not reward your dog for being afraid: Give it a treat only after it's calmed down.)

Continue this desensitization program for about 20 minutes. Repeat every day, if possible, beginning each session with the tape playing slightly softer than the loudest level your dog comfortably tolerated the day before.

Once you're able to play the tape at a volume that approximates the real noise, gradually cut back on the food rewards. Then when your dog doesn't seem to miss its customary treats, begin withdrawing your attention. Next leave the room while the tape is playing. Return after a minute or so. Gradually increase the length of your departure. One note: Each time you withdraw something else (treats or attention), you must begin the tape at a lower volume and increase it gradually.

THE FINAL TOUCHES

Once your dog reliably ignores the recording, you can add some other storm stimuli. For example, if you're using a recording of a thunderstorm, darken the room, add

Fear-Fighting Tips

Here are some additional hints for dealing with a noise-sensitive dog.

Don't cage your noise-phobic dog. Your dog may injure itself trying to get out. Use a cage only if the dog normally regards it as its safe area.

Go slowly. If you proceed too rapidly with a behavior modification program—and your dog becomes afraid during treatment—you risk making the problem worse.

Choose the right time. Plan your treatment program when it is unlikely that your dog will be exposed to the real thing. Winter is usually a good time to start: Thunderstorms and firecrackers are uncommon threats during this season.

Don't use a choke chain or other forms of punishment during the sessions. After all, the idea is to get your dog to relax.

Soundproof your dog's safe place with pillows or other sound insulation. Your dog may then go to its safe place on its own during a storm rather than being destructive.

If all else fails, consider tranquilizers. Some dogs just don't respond to sound recordings, so behavior modification won't work on these animals. For them, tranquilizers may be a good alternative. These drugs can help the dog experience a real storm without being extremely anxious. Plus through these experiences, the dog may learn to be less afraid when not tranquilized. Ask your veterinarian for the proper prescription.

"lightning" using a camera flash unit or strobe, and let it "pour" (turn on the shower). When you add the new stimuli, bring out the food treats again and start from the beginning. Although you will probably be able to proceed at a faster pace than previously, remember to avoid evoking a fear response.

When you finish the entire program, your dog should be better able to tolerate a real thunderstorm or fireworks display or another noise. If you are at home when the noise begins, take your dog to its safe place and practice the commands for food treats. (If you are not at home, make sure your dog's safe place is readily accessible.)

The treatment of noise phobias, although tedious, is usually successful. Patience is the key. If your treatment does not seem to be working, don't give up. Contact your veterinarian or a qualified animal behaviorist in your area for help.

PART • SIX

Happier Pet Ownership

· 36 ·

A Pet Owner's Guide to Worry-Free Vacations

It's a decision most pet owners face each time they contemplate going on a vacation: Should you give in to those sorrowful eyes that say "Take me along!" or ignore them and go on your way without your faithful companion? Either way, you have to plan carefully to make sure you and your furry friend have a safe, comfortable vacation.

CONSIDER YOUR OPTIONS

Although taking your pet on holiday with you may seem irresistible, I urge you to think it over carefully. Sometimes leaving your pet at home might be a better option for both of you. Answering the following questions will help you decide what's best.

Does your pet like to travel? A long car trip with a yowling cat or a carsick dog won't be much fun for either of you.

What's your mode of transportation? Except for dogs specially trained to assist blind or deaf people, most train and bus lines do not accept pets as passengers or cargo. Most airlines, however, do take pets. (The exceptions are puppies and kittens under eight weeks old.) And for some animals, travel by air can be risky. Because of a tendency toward respiratory problems, short- or flat-faced breeds such as bulldogs, pugs, Boston terriers, Pekingese and boxers often cannot handle the rapid breathing caused by travel-induced stress. Sick or ailing animals are best left behind. And older pets should get the go-ahead from your veterinarian.

What's your destination? If you plan to travel outside the continental United States, taking your pet along may be complicated. England, for example, requires an application filed six weeks before your arrival, plus a six-month quarantine in government kennels once you arrive. Even our own state of Hawaii requires a 120-day quarantine on arrival, at the owner's expense. If the quarantine lasts longer than your planned visit, what's the point?

What will your pet do when you get there? Your dog may love to go hiking and camping but probably wouldn't enjoy being cooped up in a strange hotel room all day while you are out sightseeing.

LEAVING THEM BEHIND

If you decide that your best option is to leave your pet behind, you must work out where it will stay. Many pets, especially cats, prefer to stay at home. You may be able to get a friend, relative or neighbor to feed, walk and tend to your pet. If you can't, you might want to consider one of the numerous pet-sitting services (usually listed in the Yellow Pages) that are generally reliable and insured. To play it safe, ask your veterinarian for a reference.

The alternative is to board your pet at a kennel. Before you pick one, make sure you get references and take a tour of the entire facility—including housing, feeding and exercise areas. There should be someone either on the premises or nearby at all times, with a veterinarian readily available.

TAKING PETS ALONG

If your trip just would not be complete without your four-footed family member, by all means take it along. But shortly before your trip, have your pet thoroughly examined by your veterinarian. A current health certificate, proof of vaccination and medical history are very important papers to pack.

If you will be traveling to an area where heartworm disease is a common problem, ask your veterinarian about prevention. If your cat will be visiting other cats, be sure that its vaccinations—especially for feline leukemia—are up to date. If you will be traveling to another country, check with its embassy for entry requirements.

Don't forget to pack the following items.

- A leash and collar or harness with appropriate ID tags and licenses
- All prescribed medications
- Bowls for food and water
- Food and treats
- Plenty of water
- Disposable litter boxes and litter
- Grooming tools
- Toys and a favorite pillow, bed or blanket
- A portable scratching post
- ID photos, in case your pet gets lost

If your pet will be traveling in a crate or carrier (these are required on airplanes and are the safest method of travel in cars), see that the animal has plenty of room to stand up, turn around and lie down comfortably. A week or two before you leave, start getting your pet accustomed to the crate. Begin to feed your pet in it and to include it in games, so by the time you leave, your pet will see the crate as a safe place.

Most pets do not need tranquilizers when traveling. In fact, tranquilizers can be dangerous, especially for a pet traveling by air: The medication can repress the animal's breathing, which may already be compromised by excitement and poor air quality. If your pet is extremely active, however, and you feel it travels better when mildly sedated, consult your vet. Always test tranquilizers at home beforehand to see how your pet reacts.

To help prevent travel sickness, feed a light meal no later than six hours before the trip, but provide free access to water up to the time you leave. If your pet is prone to travel sickness, you might want to ask your vet about dimenhydrinate and meclizine, two over-the-counter antihistamines that may help keep the animal from becoming ill. The dimenhydrinate dosage for cats and small dogs is 12.5 milligrams (one-fourth of a tablet) one or two times a day; for large dogs, 25 to 50 milligrams (one-half to one tablet) one or two times a day. The meclizine dosage for cats and small dogs is 12 milligrams (one-half of a tablet) once a day; for medium to large dogs, 15 milligrams (one tablet) once a day. *Always* check with your veterinarian before giving any "people" medication to your pet.

TRAVELING BY CAR OR AIR

Car travel is probably the easiest for pets because they can be with their owners. Make sure your pet is restrained, either with a travel harness designed for animals or in a crate. A cat that decides to stray over to investigate the brake pedal can be very dangerous.

Try to follow your pet's normal routine as closely as possible. Dogs should be walked as often as they usually are, and cats, offered their litter boxes as often as needed. One exception: Try to feed less than usual, as your pet will probably be less active while traveling. If car sickness is a problem, feed at the end of the day. Stop often to give your pet water if a drinking bowl is impractical in the car.

Never leave your pet inside a closed car on a warm day. The inside temperature can rise dangerously high within minutes and cause heatstroke, which is life-threatening. If you must leave your pet in the car, park in a shady area. Open the windows and sunroof enough for maximum fresh air from the outside but not enough for the pet to jump out or for someone to get in and steal the pet—or the car.

If you're planning a plane trip, check with each airline because regulations and fees vary. Most require health certificates signed by an accredited veterinarian within seven to ten days of departure.

The space allotted for animals may be limited, so reservations must be made well in advance. Some airlines allow a small pet in the passenger cabin if its cage can fit under the seat in front of you. All others travel as cargo. Either way, you need a container that is approved by the airline. These can be purchased right at the airport or from most pet stores. Print "live animal" in big letters on the top and one side of the carrier, and indicate the correct upright position.

Try to book a nonstop flight and put your pet onto the airplane at the last possible moment. Avoid travel during peak travel times and extreme weather conditions.

• 37 •

A Safe and Happy
Holiday for Your Pet

▬▬▬

The holidays are wonderful times for both people and their pets. Dogs and cats often partake in and enjoy the season's festivities as much as we do. Unfortunately, some of the things we eat or have around the house during the holidays may be dangerous to our pets. With just a few precautions, however, you can make this special time of the year safe for them.

Leave the leftovers for people, not for pets. Try to resist the temptation to have your pet sample the leftovers from your Thanksgiving or Christmas dinner. Rich foods, especially those high in fat, often trigger gastrointestinal problems in pets. Pancreatitis, a common disease of dogs, is frequently related to eating table scraps. Vomiting and abdominal pain are symptoms of this very uncomfortable and sometimes fatal condition. Sharp bones, especially those from turkey and chicken, may become lodged in your pet's mouth or throat or perforate the intestinal tract. Surgery is usually necessary to remove them. If animals don't receive immediate veterinary attention, they may die. So give your pet a rawhide or catnip gift to chew on instead of scraps.

Pass on the chocolate for your pooch. Those cute little chocolate turkeys, bunnies and Santas may be fatal to your dog! That's because dogs are especially sensitive to theobromine, a compound in chocolate. It may cause vomiting, diarrhea, heart arrhythmias, wobbliness, muscle tremors, seizures and coma, sometimes with fatal results. A 30-pound dog needs to eat only four ounces of baking chocolate to show signs of poisoning. Two or three pounds of milk chocolate would produce the same result. Cats, because of their more discriminating feeding habits, are rarely poisoned. Try to keep all chocolate out of your dog's reach. But if your dog accidentally ingests chocolate, contact your veterinarian immediately.

Secure your holiday ornaments. Your Christmas tree may bear your favorite ornaments, but it also holds some hidden dangers for your pet. Many a beautiful tree is felled by a rambunctious dog or cat. Try to support your tree securely with a sturdy stand and wires. To many dogs and cats, tinsel and ornaments seem to be terribly exciting toys. Unfortunately, those playthings often end up somewhere in the intestinal tract, causing a blockage. Try to hang your ornaments and tinsel far out of your pet's reach. If you have cats, it may be best to avoid tinsel altogether.

Safeguard electrical cords. Electrical cords also seem delectable to many pets, especially cats and young puppies. Chewed cords can cause severe burns and sometimes fatal electrical shocks. If your animal seems overly interested in electrical cords, try covering them with hot-pepper sauce or bitter-tasting commercial products sold in most pet stores.

Keep plastic wraps and ribbons out of your pet's reach. If swallowed, these, too, can cause intestinal blockage.

Protect your pet from poisonous plants. Many homes are decorated each year during the holiday season with colorful poinsettias and mistletoe. These popular plants sometimes cause problems for curious dogs, cats and birds. Poinsettias produce a milky sap that is irritating to the skin and eyes on contact and to the gastrointestinal tract if eaten. Ingestion of large amounts of mistletoe may cause nausea, vomiting and gastroenteritis. Make sure these plants are kept far out of your pet's reach. But if you do catch your pet munching on your favorite poinsettia plant or chomping a few loose mistletoe berries, contact your veterinarian immediately.

Give your pet a hideaway for the holidays. Because the holidays are traditionally times for families to get together, our homes are often crowded with people, especially young children. Many dogs and cats, even if not extremely territorial, are stressed by the increased number of strangers invading their homes. The high activity level of children is a new experience for many pets. If you feel that your pet is uncomfortable around new people, it may be best to separate it from all the intense holiday hubbub. Give your dog or cat an early Christmas present of a new bed, where it can go to get away from it all. Make sure that your young guests understand that they must let your pet rest when it is in its bed. Your cat may enjoy a new "kitty condo" or even cardboard boxes or paper bags in which to hide.

Don't let your holiday glee cause you to overlook the safety of your pet. Make every holiday season as happy for your pet as it is for you.

Gifts Fit for a King—Or a Kitty

━━━

A multitude of toys and treats that your pet would enjoy appears on the market during the holiday season. Just walk into your neighborhood pet store or supermarket for a sampling. But which of these toys and treats do pets enjoy the most? And which are the healthiest and safest?

PLAY TO YOUR PET'S INSTINCTS

Even in the wild, most mammals spend about 10 percent of their time playing. Domestic dogs and cats are no exception. Regular play sessions not only provide a fun form of exercise for your pet and keep your pet physically fit but also help deepen the bond between the two of you. From a simple ball to the most luxurious "kitty condo," toys help make playing with your pet a little more interesting for both of you. Among the variety of toys that are available, some can be used by you to interact with your pet, while others stimulate pets to play by themselves. Be sure you choose some of both.

A good game of fetch with your retriever is a very pleasant way for you to wind down after a long day, and it gives your dog a chance to have a great time, too. Tennis balls are fine, but hard rubber balls last longer. Stay away from racquet and squash balls, though, especially with larger dogs. The small size and smooth surface of these balls may cause them to get lodged in the back of the animal's throat or even to be swallowed.

Among the many inedible chew toys available, most dogs seem to prefer the cotton-rope toys, sterilized bones and cow hooves. The cotton-rope toys (Booda Bones, for example) come in many styles. Dogs like to toss them around by themselves and also play fetch with them. (Try not to let your dog get into a game of tug-of-war unless you are sure that the animal is never aggressive.)

Bones and dogs seem to go together. It's probably safest to stay away from real bones altogether, but some owners cannot resist. If you do choose to provide bones for your dog, do not give it any bone that may splinter. Rawhide bones are a good substitute. Contrary to popular belief, rawhide is nothing more than cow skin and is completely digestible. A study reported in the *Journal of the Veterinary Medical Association* demonstrated that dogs fed rawhide strips three times daily had no choking or intestinal problems. They also had cleaner teeth and did not have any undigested rawhide in their feces. Pieces of rawhide sometimes get caught in the back of a dog's mouth, however, and vomiting may occur. This rarely causes serious problems, though. If you would like to give your dog a rawhide chewy but are concerned about choking, take away the rawhide bone or strip when it is small enough to fit entirely in your dog's mouth.

Sterilized bones are inexpensive and come in several sizes. For a special treat, try filling one with cream cheese or peanut butter! Cow hooves have only recently been introduced after the many years that farm owners observed their dogs nibbling on horse and cow hoof trimmings. Hard or soft nylon bones are safe but not that interesting for a lot of dogs. One caution: Beware of squeaky toys if your dog is the type that destroys them. Veterinarians have removed far too many swallowed squeakers from dogs' intestines!

GIVE YOUR CAT THE GIFT OF EXERCISE

Cats like to chase things just as they would be chasing real prey in the wild. A simple piece of yarn pulled across the floor is great fun for most cats. Never let the cat play with string or yarn unattended, however. A string that gets lodged in the intestine may have to be surgically removed.

Many cats enjoy playing fetch with wadded-up balls of aluminum foil or other lightweight objects. And most cats love to play with the fishing rod–type toys, where you hold on to the rod or one end of a wire and your cat chases the object on the string or on the other end of the wire. Examples of these types of toys are the Kitty Dancer and the Kitty Tease.

Catnip toys or even small cloth bags filled with catnip make many cats playful. Ping-Pong balls are also favorites for cats to bat around by themselves or with you. You can even take one of your cat's toys and tie it to a doorknob with a piece of elastic to give your cat many hours of entertainment.

Curious animals that they are, cats love to play in all kinds of containers—in particular, paper bags and cardboard boxes. A kitty condo with many shelves and cubbyholes is substantially more expensive but also lasts a lot longer.

Make every holiday season fun for your pet by buying only safe toys plus tasty and healthy treats for it. Both of you will feel better as a result.

· 39 ·

On the Move with Your Cat

___

For us humans, moving to a new home is often traumatic. Packing and unpacking, leaving old friends, meeting new ones and getting adjusted to a new neighborhood can be very stressful. For cats, a move can be particularly unsettling. Unlike dogs, which are usually fine as long as they know that their owners are moving also, cats are naturally more solitary and often become very attached to their territory. In fact, although uncommon, there are stories of cats traveling long distances to their old homes after moves. Here are some suggestions to help your feline friend cope when you move.

Before the move. Do what you can to keep things the same in the old house, minimizing any sense of change. Try not to interfere with your cat's feeding area, litter box or scratching post. Either play with your cat in the packing boxes or put the boxes in a room that is off-limits to your cat.

If your cat goes outside, make sure that your neighbors are not feeding it. If there is no source of food in the old territory, it is much less likely that your cat will return there.

By all means, make sure your cat is neutered. Neutering, especially castration in males, makes cats less likely to roam.

During the move. Due to the chaos and confusion, be sure the cat remains in a safe, confined area. Movers leave doors open and do not think about pets, so a cage is probably the best place for Kitty.

Some people board their pets during moves. But unless there is no other way, I don't recommend this, since boarding is just one more change your cat has to cope with. Move your cat's carrier last, and keep your cat in it until things have settled down.

At the new home. Spraying, or territorial marking, is very common after a move, particularly among males and neutered males but also among females and spayed females. If it continues for long after the move, ask your veterinarian to prescribe tranquilizers to help stop this behavior.

Before you let your cat out of its carrier, make sure you have prepared its litter box, feeding area, bed and scratching post. The litter box should be put in a readily accessible place similar to where it was in your old home. Then move it, if you care to, after your cat is accustomed to the new home.

Choose a feeding area that is far away from the litter box and in a comfortable place close to the family. The same goes for the bed and scratching post. Show your cat where all these important places are and then let it investigate.

If your cat will be sharing its new home with other animals, make the introduction gradual. First keep the animals in separate rooms until they seem to accept each other through a closed door. Then put one in a cage or carrier and let them see each other. (If the other animal is a dog, make sure it is trained and understands commands.) Again, when they seem calm, let them be together with supervision.

If your cat will be going outdoors, try to keep it in for about two weeks. When you do finally let your cat outside, try to supervise it at first. Other cats may have included your property within their territorial boundaries and are likely to pick a fight.

• 40 •

Lost and Found Pets

―――

Of the estimated 22 million dogs and cats reported missing each year in the United States, only about 40 to 60 percent of lost dogs and 5 percent of lost cats are found, according to Penny Cistaro, manager of shelter operations at the Massachusetts Society for the Prevention of Cruelty to Animals (MSPCA). That's a shame because there is much owners can do to increase the chances of recovering lost pets; even better, there are good ways to prevent losing them in the first place. Here are some basics.

Know where your pet is at all times. Don't let your pet roam unsupervised. And take care not to let it slip out the door by accident. Mount door and gate latches beyond the reach of small children. (Too many pets escape because little kids inadvertently let them out.) Be especially vigilant if you take your pet on vacation.

Identify your pet. Of course, a dog or cat can't recite its name, address and telephone number, so it's essential that *you* supply some identification for your pet. Most shelters agree that the least expensive method, a tag and license, is still the best. Ask your vet or a pet store about having one made for your pet. Include your name, address and a day and evening telephone number on the tag. (Whether or not you include your pet's name is a matter of personal preference. Some people believe that a pet is more likely to warm up to thieves if they call it by name.) If you take your pet on vacation, add a temporary ID tag that has your vacation address.

Most towns require that dogs be licensed, and some now require licensing of cats as well. Since the license tags are inscribed with the town, state and registration number, they are excellent aids in tracing lost pets.

Another popular method of identification is the tattoo. For a small fee (usually $15 to $20), you can have your pet marked with a number provided by the tattooist, with your dog's American Kennel Club registration number or with your Social Security number. The procedure is done with a rub-on anesthetic and is painless. It takes less than five minutes.

Cats are best tattooed on their bellies. Dogs are usually tattooed on their inner thighs. Ask your vet or local shelter for referral to a reputable firm for such a service.

After your pet is tattooed, you need to register the number. This usually costs about $30. Some registries provide a tag for the animal listing both the ID number and a hotline telephone number for the registry. They also give you a copy of that information. Some companies distribute posters and offer rewards if your pet gets lost. Look for a registry that is national and has been in business for a long time. ID Pet (74 Hoyt Street, Darien, CT 06820) and National Dog Registry (P.O. Box 116, Woodstock, NY 12498-0116) are two such firms, and they register all kinds of pets.

Tattoos are helpful for discouraging theft, especially by those who sell animals for research (research facilities don't accept tattooed animals). However, tattooing does have some drawbacks if used as the only method of identification. For example, many people who find lost pets don't think of looking for tattoos, and those who do may not be able to find them. Furthermore, even if a tattoo is found, the person may not know how to contact pet registries.

A computer-age tag—a tiny microchip that is injected just beneath your pet's skin between its shoulder blades—is now available. (Farmers have been using these safe, relatively inexpensive microchips for some time to identify their livestock.) It will cost you about $20 to $30 for insertion and $30 for lifetime registration. InfoPet Identification Systems (517 West Travelers Trail, Burnsville, MN 55337) is an international firm that uses this method to identify and register pets.

But this method also has drawbacks. As with tattoos, people who find your pet need to know the microchip's there; then they must find a vet or shelter equipped with a scanner to read it! So an ID tag is still necessary.

Get your pet to say "cheese," so you can take good photos for identification. It's important to have them in case you need to describe all your pet's identifying characteristics accurately—unusual markings and color as well as sex and breed.

HAPPY RETURNS

Here's what to do if your pet does get lost despite your best efforts.

Take immediate action. Don't wait for your pet to come home on its own. The longer you wait, the more trouble your pet can get into.

Make sure it's really lost. Conduct a complete search of your premises first. Pets, especially cats and frightened animals, often hide in closets, attics and basements.

Make the right contacts. Call all area humane societies, animal shelters and animal control officers to notify them that your pet has been lost. Leave a detailed description and your phone number. Check in with them daily.

Put your pet's picture on a poster. Circulate flyers around your neighborhood and in other places your pet might visit. Make sure you include local shelters, kennels, pet stores, groomers and veterinary hospitals. If you can't include a photograph, be especially careful to provide a full description (most people do not know what all breeds look like), along with a day and evening telephone number where you can be reached. Offer a reward, but to avoid attracting dishonest people, don't state the amount.

Do some legwork. Search your neighborhood at several different times during the day. Talk to children, mail carriers, shopkeepers—anyone who spends a lot of time outside. Cats usually stay close to home (within a 2-mile radius, according to the MSPCA), but dogs may roam as far as 20 miles.

Put out the welcome mat. To try to attract your pet, place an article of your clothing and some food in a box outside the door the animal usually uses.

Launch an ad campaign. Place ads in the classified sections of your local newspapers and check the lost-and-found columns in those papers as well. If you think your pet may have been stolen, check the pets-for-sale columns, too.

Make broadcast news. Call local radio stations. Some announce lost-pet information.

Call in the feds. The U.S. Department of Agriculture agent in your area can give you the names of dealers that sell stray animals for research. Leave a description of your pet and your phone number with every one you contact.

Be aggressive in protecting and retrieving your pet. It is the only way to fight against a threat all pet owners must face.

PART SEVEN

Selecting the Breed That's Right for You

· 41 ·

A Guide to Popular
Dogs and Cats

After deciding that you are ready to take on the responsibility and commitment of owning a pet, your next step is to consider carefully which type of dog or cat you would like to have. Be sure to read chapter 1 for basic advice on choosing a pet. Start out by picking a pet that best fits into your lifestyle, for that is one of the most important factors in determining the success of the relationship you have with your animal companion.

PEDIGREES DON'T
GUARANTEE PERFECT PETS

If you decide to get a purebred animal, you should know that not all breeds suit everyone and not all individual animals within a breed are the same. A breed that is right for some people can be wrong for others.

Breeders have spent years trying to perfect not only the appearance of each breed but the breed's behavioral tendencies as well. Golden retrievers, for example, were bred not only for their large size, long golden coat and floppy ears but also for their hunting ability and friendliness toward people. Similarly, Abyssinian cats were bred for their "wildcat" look, intelligence and care with children.

Unfortunately, this quest for the perfect pet in terms of desired traits has led to breeding animals with similar genes, and that has produced undesirable traits in some individuals within the breed. For instance, many breeds are plagued with inherited hip dysplasia, kidney problems and eye ailments. Others are fraught with behavioral problems.

DON'T IGNORE THE MIXED BREEDS

As you will see in the guide that follows, just about every type of purebred dog can be saddled with some genetic baggage. So don't make the required emotional and financial investment in a purebred dog if it's just to ensure having a healthy pet with a good disposition. You may not get one.

That is not to say a mixed-breed dog is sure to be free of behavioral quirks, psychological peculiarities and physical weaknesses. But chances are good that the no-pedigree pup you pick out at the local shelter, usually at a much lower cost, will be as hardy as the $600 darling available from a kennel.

Adopting a pet from a shelter has two major advantages: Dogs slated for euthanasia after a limited stay (usually a month) are saved; and the demand for pets is partly satisfied by existing animals rather than additions to the overpopulation of dogs.

So if you'd be happy with a less than pure collie, schnauzer, poodle or other mix, you might get the dog of your dreams at the shelter. Or maybe what you want even more is a friendly companion—not too big, not too small—for company and protection. In either case, a mixed-breed pup might meet your needs and save you considerable money and concern in the bargain. It's worth looking into.

WHERE TO FIND THE BEST PUREBRED PET

If you do decide to buy a purebred animal, the best source is a reputable professional breeder. These specialists know all about the problems of a particular breed and are more likely to work against perpetuating undesirable traits in their breeding practices. Hobby breeders, or so-called backyard breeders, who breed "for the experience," "to make a little extra money" or because their dog or cat "should have at least one litter," are less likely to be concerned about undesirable traits that can be passed on in the offspring. And pet stores that are supplied by puppy mills usually have no information about the temperament or medical problems of an animal's relatives.

Ask a local veterinarian for the name of a good breeder, or call a local breed club for that information. Dog and cat shows are good places to view individual animals of all breeds and to interview some breeders. Don't look to the American Kennel Club and United Kennel Clubs for guidance in your search, for they do nothing more than register dogs (although the former can help you locate breed rescue leagues). And the numerous cat registries (the Cat Fanciers Association is the largest) have the same limited function. These organizations do not vouch for the quality of the dog or cat that has been registered.

A pedigree ("papers") merely proves a pet's ancestry. It says nothing about an animal's quality. Your potential pet's family members could all have fancy names, but each one might be a dog or cat with health or behavior problems.

Several good books are available on dog and cat breeds. My favorites are *Your Purebred Puppy* by Michele Lowell and *The Perfect Puppy* by Benjamin and Lynette Hart. For cats, I like *The Cornell Book of Cats,* edited by Mordecai Siegal.

The information in the guide that follows is a combination of many experts' opinions on breeds, mixed with my personal experience as an animal behaviorist. Perhaps some of my own opinions are a bit slanted, since the animals I see professionally all have behavior problems, but I have made sure to take others' views into account as well.

Because of space limitations, only the most popular breeds of cats and dogs are listed. Those that don't appear—collie, Scottie, bulldog and Great Dane, for example—may be out of fashion just now. Your local library can probably furnish full information on any breed not covered here.

Dogs

Basset Hound

Europe has several varieties of basset, including some that are taller and have straight legs. The popular American basset, standing up to 15 inches at the shoulders, has short, crooked legs and a long, heavy body, weighing 40 to 55 pounds. The coat is short, and the skin is very loose. Only a quick brushing once a week is needed, but frequent baths may be required because this breed tends to have a strong odor. Shedding is moderate. Bassets are seen in combinations of black, white and/or tan. The ears are heavy and hang down; when pulled forward, they extend beyond the nose. The tail is long.

Origin: Bassets are scent hounds descended from the French bloodhound and the St. Hubert hound; the name *basset* comes from the French *bas,* meaning "low to the ground." Raised by royalty, chiefly in France and Belgium, these dogs were used for moderately slow trailing of deer, hare, rabbit and other game. In the United States, the basset soon earned respect for its superior scenting ability and became a popular hunting companion. Its popularity as a pet is fairly new.

Personality: Generally very mild-mannered, some bassets are dignified, and others are clownish. They require daily exercise to stay in shape, as they tend to put on weight. Their stubborn nature requires patience when it comes to obedience training. (Only gentle techniques should be used.) Housetraining may also be a bit slow. Most bassets are friendly to strangers, but some are reserved. Individual males may be dominant, guarding their food and toys and growling or snapping when punished or ordered to do something they don't want to do. Females seldom have these problems. Bassets are not my choice of breed for novice owners.

Choose a female if there are children in the house.

Potential medical problems: The body shape of bassets makes them susceptible to bloat and vertebral disk problems; also look out for chronic ear infections, glaucoma, eyelid abnormalities and skin diseases.

Beagle

These dogs come in two size varieties: standing 10 to 13 inches at the shoulders and weighing 18 to 20 pounds, and standing 13 to 15 inches at the shoulders and weighing 20 to 30 pounds. The short and hard coat sheds in moderate amounts, so only a quick brushing once a week is required. Beagles are seen in combinations of black, white and/or tan. The ears hang down and, when pulled forward, nearly reach the nose. Beagles are known for their characteristic baying bark.

Origin: One of England's oldest breeds of hound, beagles were originally bred to hunt hare, and they run in packs. The name comes from a group of hunting hounds called begles.

Personality: Good-natured, playful and gentle dogs that require a moderate amount of exercise. They do well as city dogs if given brisk daily walks and an occasional run. Most beagles are friendly to strangers, but some tend to be timid, so exposure to people and new places at a young age is especially important. Obedience training may be difficult because beagles tend to be distracted easily, but housetraining is usually simple. They tend to bark and howl at sounds and things going on outside, and some do so excessively when left alone. A good breed for novice owners, and very good with children.

Potential medical problems: These pups are often troubled by heart disease, vertebral disk problems, bleeding disorders, epilepsy, skin conditions, glaucoma and cataracts. Beagles also love to eat, and they tend toward obesity.

Boxer

A medium to large, muscular working breed, the boxer stands 21 to 25 inches at the shoulders and weighs 60 to 75 pounds. The short coat needs only a quick brushing once a week; shedding is only moderate. Boxers are seen in fawn (light tan), mahogany brown or brindle (brownish with black stripes). White markings on the face, chest, neck and feet are common. The square muzzle is usually black or black and white. The ears must be cropped for show but are often left hanging in pets. This is one of the few breeds whose teeth do not meet naturally. The lower jaw protrudes out in front of the upper jaw. The tail is docked.

Origin: Descended from sixteenth-century European bulldogs, the boxer was developed in Germany by crossing with a terrier strain. Although first used as a dogfighter and bullbaiter, the boxer later gained popularity as a police dog and companion. The name *boxer* comes from the dog's manner of using its front paws somewhat like a human boxer whenever fighting or playing.

Personality: A very nice, good-natured dog. Active and playful with a good sense of humor, boxers have been called "honest" dogs because their faces express their emotions. They adapt well to city life if given sufficient exercise. Vigorous games and brisk walks are always appreciated. Most boxers are very friendly to strangers, but some are extremely protective. They may show aggressiveness to strange dogs. Boxers respond well to patient and gentle but firm obedience training. Housetraining comes easily. They are good dogs for novice owners and good with children. Boxers are excellent family dogs.

Potential medical problems: Unfortunately, boxers are a short-lived breed, rarely living past age ten. They are susceptible to cancer, digestive problems, heart disease, corneal problems and bloat.

Chihuahua

These long-legged toy dogs stand about five inches at the shoulders and weigh between two and six pounds. They are seen in two coat types: The smooth shorthair (which needs only a quick brushing once weekly) and the longhair (which has longer hair on the ears, chest, stomach, legs and tail that needs to be brushed and combed every other day). Shedding is moderate. Common colors include white, blonde, fawn, black, black and tan and patched, although any color is acceptable. The ears are pricked, although the tips may droop slightly. The tail is long, and the head is dome-shaped.

Origin: This breed can be traced back to the ninth century A.D. as the techichi dog of Mexico's Toltec Indians. The modern version is said to have resulted from crossing the techichi with a small hairless dog from Asia. The earliest specimens of the modern breed were found in the state of Chihuahua in Mexico, for which they were named. The Aztecs, who conquered the Toltecs, revered these little dogs as religious symbols of good fortune.

Personality: Intelligent and alert, this dog is the ultimate companion for apartment dwellers. Exercise requirements are minimal; many don't even like the outdoors. Most Chihuahuas are very playful, curious and lively; some are bold and temperamental, while others are timid and nervous. They may snap at strangers if they feel threatened. Generally, these dogs do well with obedience training, except for the ones that get too nervous around other dogs. (This breed is said to be clannish, preferring its own kind to dogs of other breeds.) Housetraining can be difficult, and many owners opt to paper-train. Chihuahuas are fine dogs for novice owners, good with older people and considerate of children, especially if they grow up together.

Potential medical problems: These dogs live a long time; 15 years is not unusual. They are susceptible to dislocating kneecaps, fractures, jawbone disorders, arthritis, eye problems, heart disease, tooth and gum disease and hydrocephalus (water on the brain).

Chow Chow

Medium in size, these dogs stand 18 to 20 inches at the shoulders and weigh 50 to 70 pounds. The coarse, dense coat needs to be brushed at least every other day. The most common colors are red and black, but cream, cinnamon and blue Chows are also seen. The ears are pricked, the tail curls, and the tongue is characteristically a blue-black color.
Origin: One of the oldest recognizable breeds of dog, the Chow was developed in China as a hunter of wolf and bear. The name *Chow Chow* derived from the eighteenth-century Pidgin English term for particles from the Orient (knick-knacks or bric-a-brac, including curios and "mixed pickles") brought to England by sailors. The sailors simply wrote "chow chow" rather than describe each thing. Since the dogs often accompanied the other items, they became known as Chow Chows.
Personality: Unlike other breeds, Chows tend to be independent, introverted and serious, and although some are affectionate and friendly, most are not. They also tend to be one-person dogs with an aggressively protective and territorial nature. But as most Chows do not seem interested in pleasing their owners, they are not easily obedience-trained. Harsh training should be avoided, as many respond aggressively. Housetraining comes easily to most Chows. This is a breed for experienced owners only. Not recommended for families with children. Chows are among the cutest puppies, and many inexperienced people are unpleasantly surprised when their cute teddy bear puppies grow into aggressive adults. Ongoing exercise, socialization, supervision and careful management are essential with this breed. Because

of their thick coats, Chows are bothered by the heat.

Potential medical problems: As with so many breeds, hip dysplasia, skin conditions and eyelid abnormalities are special dangers.

Cocker Spaniel

Small in size and standing 13 to 15 inches at the shoulders, these dogs weigh approximately 25 pounds. With their silky coats and medium shedding, cocker spaniels require frequent brushing and professional grooming. May be buff, black, tricolor (black, tan and white) or parti-color (black and tan or white and tan). These dogs have very heavy ears. Field types have shorter coats.

Origin: Cockers were first bred in England as hunters, to flush and retrieve birds. (The name comes from the woodcock bird.) But very few of these dogs are used for hunting nowadays.

Personality: Because of this breed's popularity, several types of cockers are available, so be very careful when choosing an individual. Most cockers are friendly to strangers and good with children, but some, especially males, are aggressive toward their owners. Certain cockers are afraid and snappy with strangers. Many urinate submissively when greeted. In general, cockers require a moderate amount of exercise and are easy to housetrain.

Potential medical problems: Be on the alert for signs of ear infections, skin conditions, progressive retinal atrophy (PRA), cataracts, glaucoma, eyelid and eyelash abnormalities, hemophilia, heart defects and epilepsy.

Dachshund

Dachshunds are seen in two sizes and three coat types. All have characteristically long bodies and short legs.

Miniature varieties stand about five inches at the shoulders and weigh under ten pounds. Standard varieties stand about nine inches at the shoulders and weigh over ten pounds. The three coat varieties are smooth, longhair and wirehair. The smooth, or shorthairs, are the most frequently seen. Their short, shiny coats need only a quick brushing once a week and shed moderate amounts of hair. The longhairs have silky coats with longer hair on the chest, stomach, legs and tail. They require brushing and combing every other day. The wirehairs have rough coats, bushy eyebrows and beards. They also need to be brushed and combed every other day. Dead hair needs to be removed by stripping twice a year. The most common coat colors are solid red (actually brownish) and black and tan. Dachshunds are also seen in a variety of other coat colors: tan, chocolate, brown, dappled (one color background with irregular patches of other colors) and striped or brindle. All types have hanging ears and long tails.

Origin: Traced to a 35-pound hunter of badger in medieval Europe, the name *dachshund* literally means "badger dog." They were bred down in size to hunt smaller prey such as rabbit, primarily in Germany. Although still used for hunting in Europe, they are primarily companions in the United States.

Personality: Curious, bold and playful, with a good sense of humor. Both an affectionate and a responsive companion, and an alert watchdog with a loud bark. With patience and gentleness, obedience training is possible and successful. Housetraining can be a little more difficult. Harsh training techniques may cause excessive submissiveness. Dachshunds are barkers and diggers, which is not surprising for an ex–badger chaser. Most are friendly to strangers, but some will snap defensively if people move too quickly toward them. They have a lower exercise requirement than other breeds and are excellent apartment dogs. They are fine for novice owners and good with considerate children.

Potential medical problems: One of the longer-living breeds;

15-year-old dachshunds are not extraordinary. They are prone to vertebral disk problems, obesity, diabetes, urinary stones, eye disorders, skin conditions and heart disease.

Dalmatian

A medium to large, muscular dog standing 19 to 24 inches at the shoulders and weighing 46 to 65 pounds. Shedding moderately year-round, the short coat needs to be brushed several times a week. White with black or liver spots, Dalmatians are some of the flashiest dogs around. Puppies are born all white and develop spots at two to six weeks of age. The ears hang down, and the tail is long.

Origin: An ancient spotted dog, the Dalmatian has come through many generations unchanged. The actual ancestry of the breed is controversial, but the name comes from the Austrian region of Dalmatia and dates back to the middle of the eighteenth century. No breed has been involved in as many varied activities: the original coach dog, a dog of war, a draft dog, a shepherd, a hunter of vermin, a bird dog, a trail hound, a circus performer and, of course, a firefighter's mascot. Today, Dalmatians are mostly companions and are quite popular due to the Disney movie.

Personality: A hardy, high-spirited and playful dog that needs a lot of vigorous exercise. Some are too energetic to adapt well to city life. While one Dalmatian may be very friendly to strangers, another may be reserved or overly timid and may snap if approached too quickly. Some are very protective, while others, especially males, are dominant and aggressive toward their owners. Because of their high energy level, Dalmatians are easily distracted, which can make obedience training difficult. Patience and a gentle but firm method usually results in success. Housetraining is usually not a problem. Dalmatians are not a good choice for inexperienced owners and behave unpredictably around children.

Potential medical problems: These popular dogs often fall victim to hip dysplasia, inherited deafness, skin conditions and urinary stones.

Doberman Pinscher

This medium to large, muscular working dog is a very graceful-looking animal. The average Dobie stands 24 to 28 inches at the shoulders and weighs 60 to 85 pounds. Its smooth, soft coat requires only a quick weekly brushing, and shedding is moderate. Most are black or red with rust markings above the eyes, below the tail and on the muzzle, throat, chest, legs and feet. They are also seen in fawn or steel blue with the same rust markings. The ears are cropped or left hanging. The tail is docked.

Origin: Louis Doberman developed this breed in Apolda, Germany, around 1890 from a combination of breeds, including old short-haired shepherd dog stock, Rottweiler, black-and-tan terrier and smooth-haired German pinscher. At first, the Doberman (as it is called in Germany) was used almost exclusively as a guard dog. Later it was a police and war dog. Now it is also a favorite friend and guardian for many families.

Personality: People tend to think of Dobermans as "mean," but in fact, they are usually very sweet and mellow. However, some are overly energetic, bold and aggressive, while others are nervous or suspicious. Dobermans require a moderate amount of vigorous exercise and long, brisk walks. They tend to be busy dogs and demand a lot of attention. If a firm but gentle technique is used, they are easy to obedience-train and housetrain. Since Dobies were originally bred for protection, they make excellent watchdogs. Individually, they may be either reserved or overly friendly to strangers. A Doberman that was bred to be a pet, rather than to protect, is a fine choice for a novice owner. Most are good with children, especially if raised with them.

Potential medical problems: Dobermans unfortunately have been stricken with more than their fair share of medical problems. They are susceptible to hip dysplasia, von Willebrand's disease (a blood-clotting disorder), bloat, skin diseases, immune-mediated diseases, severe heart disease (cardiomyopathy), thyroid and fatal liver disorders and a vertebral disorder commonly known as wobbles.

English Springer Spaniel

Graceful, medium-size dogs, springers stand 19 to 20 inches at the shoulders and weigh 45 to 55 pounds. The coat can be straight or wavy, with longer hair on the chest, stomach, legs and tail. Brushing and combing is required every other day, with scissoring and clipping necessary every two to three months. Springers are usually seen in either black and white or liver and white. Tricolors (black, tan and white or liver, tan and white) and roans (speckled mixtures of blue or liver and white) are less common. The ears hang down and are heavy. The tail is docked. Field springers (those bred for hunting) are smaller and shorter-coated and have patches or spots of color. Show springers have concentrated areas of color and are flashier looking.

Origin: Developed in England, the springer earned its name from the dog's duties of "springing" the pheasant into the air. One of the most popular hunting breeds in the United States, the springer is pro at flushing birds from their hiding places and then retrieving them after the hunter shoots. They have recently become popular as pets as well.

Personality: Springers are usually happy, playful and high-spirited dogs. They require moderate amounts of exercise, and if given what they need, they generally adapt well to city life. Daily walks and vigorous ball-playing sessions are usually adequate. Springers love obedience training and are quite good at it. They are also easily housetrained. On the whole, springers are good family dogs, even with novice owners and with children.

Recently, however, a serious behavior problem known as "springer rage" has cropped up in many lines. Dogs afflicted with the problem tend to be very dominant. They guard food and objects and growl, snap or bite when reprimanded or told to do something they do not want to do. Some will even bite when petted if they do not want to be petted. They are unpredictable, moody and not easy to live with. The problem is hard to detect in young puppies and develops sometime during the dog's first two years. It is seen in both males and females, but mostly in males. If you do decide to get a springer, make sure you examine your potential puppy's bloodlines carefully for any evidence of this disorder.

Potential medical problems: This breed is susceptible to progressive retinal atrophy (PRA), hip dysplasia, eyelid and retinal abnormalities, ear infections and skin conditions.

German Shepherd

An agile but rugged medium-size dog, the German shepherd stands 22 to 26 inches at the shoulders and weighs between 65 and 100 pounds. The coat, which is harsh with a dense undercoat, sheds abundantly year-round and must be brushed several times weekly. The ears are pricked, and the tail long. The most common color is black and tan, although bicolors (black with tan points), golden sables (golden tan with black-tipped hairs), gray sables (gray with black-tipped hairs) and solid blacks are also seen. White is not accepted by the American Kennel Club but is by other clubs.

Origin: Bred as a sheepherding and farm dog in Germany, this breed has been developed both structurally and temperamentally so that shepherds are now among the best guard, police and military dogs. They are used extensively for search and rescue, for narcotics and bomb detection and as guide dogs for the blind.

Personality: Very intelligent and responsive to obedience training. Easy to housetrain. Requires a medium amount of physical exercise, but also needs mental exercise. Many shepherds are excessively protective of owners and property and require firm management. Some are overly shy and develop into fear biters. It's worth the time it takes to choose a breeder who strives for good temperament, health and, preferably, working ability in obedience or herding, not in protection work. Training and exposure to all kinds of people should begin during puppyhood. This is not a breed for the inexperienced.

Potential medical problems: Be alert for hip dysplasia, bloat, digestive disorders, bone ailments (panosteitis), spinal problems, eye disease (pannus) and skin conditions.

Golden Retriever

A powerful and active medium-size dog standing 23 to 26 inches at the shoulders, weighing 55 to 75 pounds and possessing a kindly expression. The coat is more than 2 inches long over the body, with longer hair on the chest, stomach, legs and tail. Shedding is moderate, requiring brushing at least twice weekly. Seen in a variety of "golden" colors, from dark red to medium gold and light cream. The ears hang down.

Origin: The breed originated in England as a duck retriever, which worked well both on land and in water. Experts believe it derived from crosses between setter, water spaniel and other water breeds and the lightly built St. John's Newfoundland (also the predecessor of the Labrador and flat-coated and curly-coated retrievers). Now, because of the golden retriever's popularity as a pet, most of its hunting skills have been lost.

Personality: The breed's cheerful demeanor and outgoing, demonstrative spirit have made the golden one of the world's favorite family dogs. Eager to please, goldens take well to obedience training and are easy to housetrain. Also one of the friendliest breeds, some are overexuberant and jump on people when greeting them. This breed requires daily vigorous exercise. Some individuals are timid, and a few are hyperactive. Many are afraid of loud noises, such as thunder and fireworks. Recently, because of the golden's popularity, aggression problems have cropped up. Extra care should be taken to find a responsible breeder and meet the pup's parents.

Potential medical problems: Be vigilant for hip dysplasia, progressive retinal atrophy (PRA), von Willebrand's disease (a blood-clotting disorder), heart problems, cataracts, skin conditions, eye abnormalities and epilepsy.

Labrador Retriever

A medium-size, strongly built dog standing approximately 22 to 25 inches at the shoulders and weighing 55 to 75 pounds. The coat is short, very dense and straight, shedding a medium amount of short, stiff hair. Brushing once weekly is sufficient. The ears hang; the tail is long and otterlike (round and thicker at the base, with no feathering). Seen in black, yellow (golden or cream) or chocolate (dark brown).

Origin: In the province of Newfoundland in Canada, this type of dog was first developed as a water retriever. The breed was further developed in Britain, where it received the name *Labrador*. In addition to their popularity as pets, Labs are among the best hunting and field trial dogs.

Personality: One of the nicest family dogs, Labs require a good deal of vigorous daily exercise and play; they love to play ball and chase sticks. They need to be around people. Some Labs become very destructive if underexercised and left alone for long periods. Very easily obedience-trained and housetrained. If its needs are met, this dog adapts well to city life. Some are overactive, and some (especially males) are aggressive, so be careful in choosing an individual. Both field types and show types are available.

Potential medical problems: Watch out for hip dysplasia, elbow dysplasia, progressive retinal atrophy (PRA), cataracts, eyelid and retinal abnormalities, bloat.

Lhasa Apso

These are small dogs, standing 9 to 11 inches at the shoulders and weighing between 13 and 15 pounds. The coat is long, straight and soft, requiring daily brushing and combing. The head, including the face, is overhung with hair. Shedding is minimal with proper grooming. Many Lhasa owners have the dog's coat trimmed short during the warm months. The ears hang down, and the tail is heavily feathered and curls over the back. The mouth may be even, although many have lower jaws that protrude.

Origin: In Tibet, its homeland, this breed is known as *abso deng kye,* the "bark lion sentinel dog." Kept as special guards inside dwellings, they bark to warn inhabitants of approaching intruders. Along with the fierce and powerful Tibetan mastiff, they are kept chained to a post beside the outer door. Together, these two breeds protect their masters and their homes.

Personality: Playful, assertive, bold and independent. Although they are among the cutest dogs, most do not like to be handled excessively and are not cuddly lapdogs. They are stubborn and difficult to obedience-train but easy to housetrain. Harsh training techniques should be avoided, as many become aggressive. Lhasas tend to be suspicious of strangers, since they were bred to be guard dogs. They don't require much exercise, but obedience training, patience and proper management are essential. Many of these dogs, especially males, guard their food and toys and snap if disturbed while resting, if punished or even if urged to do something

they don't want to do. Lhasas are for experienced owners only and are usually good with older, considerate children.
Potential medical problems: Keep a keen eye out for kidney problems, skin conditions and eye trouble.

Miniature Schnauzer

A small dog standing 12 to 14 inches at the shoulders and weighing 13 to 15 pounds. The coat is short and wiry, with bushy eyebrows and beard. Brushing and combing is required about twice weekly. Trimming or stripping is required every one to three months. Shedding is minimal. Most are seen in salt and pepper (gray with light- and dark-banded hairs, with silver muzzle, chest and legs), although blacks and black and silvers also exist. The ears are either cropped or naturally flop forward. The tail is docked short.
Origin: All three schnauzers—miniature, standard and giant—originated in Germany, but they are separate breeds. The modern standard schnauzer was probably derived from the crossing of black poodle and wolf-gray spitz with old German pinscher stock. They were yard and stable dogs used for their guarding and ratting abilities. The miniature was produced by selection of small size and crossing with the affenpinscher. By type and temperament, they resemble the British terrier. And although they are classed as terriers in the United States, they are not considered as such in Europe.
Personality: Hearty, intelligent and active dogs that are both lovable and playful. With a terrier spirit, schnauzers need a moderate amount of exercise, and they like their walks. While most are friendly to strangers, some are reserved or timid and tend to snap if approached too quickly. Most bark furiously when people walk by or enter the house, even if they are friendly. This tendency to bark makes them good watchdogs. They usually do well with gentle obedience training and are easily housetrained. They are a good

choice for apartment dwellers, if they get the chance to get out and exercise. Also a good choice for novices and for families with considerate children.

Potential medical problems: Be aware of schnauzers' susceptibility to cataracts and corneal disorders, bladder stones, bleeding disorders, liver disease, heart disease, diabetes, skin conditions and cysts known as "Schnauzer bumps."

Pekingese

Classified as toy dogs, Pekes stand eight to nine inches at the shoulders and weigh up to 14 pounds. The coat is coarse, with a thick ruff and profuse feathering on the ears, chest, legs and tail. These dogs need to be brushed and combed every other day. Shedding is moderate. They are seen in red, fawn, black, white, black and tan, patched or parti-color, sable (brown with black-tipped hairs) and brindle (brownish with black stripes and flecks). The face is always black with spectacles around the eyes. The ears hang down, and the face is "pushed in." The heavily feathered tail is curled over the back.

Origin: In China, during the Tang dynasty of the eighth century, these dogs were considered sacred. Kept as pets by the imperial families, they were given the names *lion dog* and *sun dog* (because of their thick, richly colored coats) and *sleeve dog* (for those carried in the sleeves of imperial household members). When the Imperial Palace at Peking was looted by the British in 1860, the Pekingese was introduced into the Western world.

Personality: A dignified, independent and regal dog, it has a streak of stubbornness that some owners find endearing. The breed has minimal exercise requirements, is calm indoors and is usually not noisy. A Pekingese is an excellent choice for apartment dwellers. Reserved with strangers, they are generally one-family dogs. Though easy to house-train, few are interested in obedience training but are usually well mannered right from the beginning. These are good dogs for novice owners, and they do well with older, considerate children, especially if raised together. Also, Pekes are excellent companions for elderly dog lovers.

Potential medical problems: Because of their bulging eyes, Pekingese are susceptible to corneal problems and prolapsed eyes, in which trauma causes the eyes to come out of their sockets. Eyelid/eyelash abnormalities are common. Pekingese suffer from respiratory difficulties and heatstroke, plus collapsed nostrils, cleft palate and jaw disorders, because of their pushed-in faces. Vertebral disk problems and urinary bladder stones often occur.

Pomeranian

A fluffy toy dog that stands six to seven inches at the shoulders and weighs three to seven pounds, the Pomeranian has a long coat that is soft and thick, with a profuse neck ruff. Poms require daily brushing, and shedding can be significant if this is not done. The solid colors—red, white, black, brown, blue, orange and cream—are most commonly seen, sometimes with black shadings. Particolors (white with color patches), sables (tannish gold with black tipped hairs) and black and tans are less common. The ears prick up, and the profusely feathered tail is carried over the back.

Origin: The Pomeranian is descended from the sled dogs of Iceland and Lapland. In Pomerania, Germany, its larger predecessor was an able sheepherder. Over the years, starting in Pomerania, the dog has been bred down in size.

Personality: A vivacious and spirited little dog, the Pomeranian is said to be one of the sturdiest of the toys. An ideal apartment dog, as most exercise requirements can be fulfilled either inside or on daily walks. Most respond well to gentle obedience training and are easy to housetrain. Generally reserved with strangers, most warm up once they get acquainted. Some are excessively timid and yappy and may snap if approached too quickly. Poms are a good choice for novice owners and are good with considerate children, especially if they grow up with them.

Potential medical problems: Guard against dislocating kneecaps, eye infections, collapsing windpipes, heart disease, skin conditions and tooth and gum disorders.

Poodle

The standard measures 22 to 26 inches at the shoulders and weighs 45 to 60 pounds. The miniature stands 10 to 15 inches at the shoulders and weighs 14 to 17 pounds. The toy is no taller than 10 inches at the shoulders and weighs only 5 to 7 pounds. All have curly coats that need frequent brushing and professional grooming. But they hardly shed, which makes this breed an excellent choice for allergic people. The ears hang down, and the tail is docked. Although parti-colors are seen, only solid colors—black, white, gray, blue, silver, cream, red, apricot, chocolate (dark brown) and cafe au lait (light brown)—may be shown.

Origin: The standard poodle originated in Germany or Russia as a duck retriever. The name *poodle* comes from the German word *pudelin,* which means "to splash in the water." France was responsible for using the breed as a show and circus dog, which gave rise to the common reference "French poodle." The standard was bred down to the miniature and toy varieties.

Personality: Very intelligent and easy to housetrain. Poodles are usually very lively and playful, but some, especially toys, are high-strung and timid with strangers. Others, primarily miniatures and toys, are excessive barkers. Generally good with children and other animals. Poodles require a medium amount of exercise.

Potential medical problems: Check for special concerns in each size. *Standard:* Hip dysplasia, progressive retinal atrophy (PRA), von Willebrand's disease (a blood-clotting disorder), bloat, cataracts and skin disease. *Miniature:* PRA, cataracts, glaucoma, eye and ear infections, digestive problems, skin conditions, heart disease and epilepsy. *Toy:* PRA, cataracts, glaucoma, ear infections, digestive problems, skin conditions, diabetes, heart disease, epilepsy and dislocating kneecaps.

Rottweiler

These are large, stocky dogs, standing 24 to 27 inches at the shoulders and weighing at least 85 to 115 pounds. The coat is coarse, dense and short and requires once-a-week brushing. Shedding occurs but is not excessive. All are black with rust markings above each eye, below the tail and on the muzzle, throat, chest, legs and feet. The ears hang down, and the tail is docked short.

Origin: Developed in the town of Rottweil, Germany, from a large, short-haired Roman cattle dog. The butchers and cattle merchants who settled in Rottweil kept these dogs as herders, companions and guard dogs. It was customary for a master to tie his purse around his dog's neck to protect the money. As donkeys and trains replaced dogs for moving cattle and goods, the breed was almost lost. A renaissance for the Rottweiler began in 1910, when it was found to be a desirable breed for police training.

Personality: A calm, confident, serious dog that is usually even-tempered. A powerful breed and excellent guard dog, Rottweilers are usually reserved and not overly friendly to strangers. A medium amount of exercise is required. Intense socialization, obedience training and careful management and supervision are essential. This breed is for experienced owners only. Many Rottweilers, especially males, tend to be excessively dominant to their owners and aggressive to other people. Housetraining comes easily, but this breed is not my recommendation for families with children. Pets should not be obtained from parents that were bred for protection work. **Potential medical problems:** Look out for hip dysplasia, bloat and retinal and spinal cord problems.

Shar Pei

A "round-looking" dog standing 18 to 20 inches at the shoulders and weighing 40 to 55 pounds. The harsh coat may be smooth or 1 inch long, and a quick brushing once weekly is all that is needed. Shedding is moderate. Only solid colors—black, cream, fawn, chocolate and red—are accepted. Both heavily wrinkled, massive-headed dogs and tighter-skinned, smaller-headed dogs are seen. The tiny ears fold forward. The tongue is blue-black like a Chow's, and the tail is ringed and held high.

Origin: This very old Chinese fighting dog, whose loose skin and folds enable twisting without injury during a fight, was once nearly extinct. Now the breed is quite popular and ready for full American Kennel Club recognition.

Personality: A quiet, serious and reserved dog. Most are not friendly to strangers. Although independent, most Shar Peis are affectionate to their owners. Requiring relatively little exercise and known for being very clean dogs, Shar Peis adapt well to city life if they get some exercise and proper supervision. Many are aggressive to other dogs and prey, so they must be kept on leashes, except in confined areas. Obedience training can be difficult, but most will respond if the method is gentle but firm. Shar Peis are easily house-trained. They are neither dogs for inexperienced owners nor the best dogs for households with children.

Potential medical problems: Owners would be wise to set aside a separate bank account for dealing with their Shar Pei's medical problems. Among the common ones are hip dysplasia, eyelid abnormalities, ear infections and chronic skin conditions.

Shetland Sheepdog

This small dog stands 13 to 16 inches at the shoulders and weighs 14 to 18 pounds. The beautiful long, thick coat must be brushed and combed every other day, daily when shedding. Shedding occurs in large amounts. Shelties are seen in sable (gold to mahogany with white markings), tricolor (black with white markings and tan shadings) and blue merle (mottled blue, gray and black with white and/or tan markings). The ears are three-quarters pricked, with the top quarter tipping forward.

Origin: This small sheepherder hails from the Shetland islands near Scotland. Because of the rugged environment, along with scarce food and space, many of the animals produced on the Shetlands are miniature in size. Although the present-day large collie has played a small part in the development of today's Sheltie, the original working Sheltie was bred from a black and white, medium-size collie (like today's Border collies). Shetland sheepdogs are a breed unto themselves. They are not miniature or toy collies.

Personality: Overall, an extremely bright, gentle and responsive breed. They are among the easiest of dogs to obedience-train, as it is their nature to obey willingly and happily. Housetraining is also simple. Many are high-strung and bark excessively, however. Some also circle and nip at heels in an attempt to herd. They need a moderate amount of daily activity and do well in apartments if given sufficient exercise. They are among the nicest breeds for families with children and are a good choice for novice owners.

Potential medical problems: These sweet dogs are susceptible to progressive retinal atrophy (PRA), collie eye anomaly, heart disease, epilepsy and, in blue merles, deafness.

Shih Tzu

This toy dog, which stands 8 to 11 inches at the shoulders and weighs 9 to 18 pounds, has a coat that is long and soft and needs to be brushed and combed every day. Shedding is minimal with proper grooming. All colors are seen, including multicolored varieties. The ears hang down and are covered with long hair. The teeth may meet in a level or undershot bite, where the lower jaw protrudes. Noses are short, and faces "pushed in." The tail is long and heavily feathered and curls over the back.

Origin: These were the cherished dogs of the Tang and Ming dynasties of China. During the Chinese revolution, many families killed their dogs rather than let them be captured by enemies. Some that survived were taken to England and then to the United States. *Shih Tzu* is Chinese.

Personality: A delightful little dog: proud, assertive and playful. Requiring little exercise, these dogs would prefer to sit in your lap. A true lover of companionship, attention and comfort. Most respond well to gentle obedience training but can be difficult to housetrain, so some people opt to paper-train. Although most are friendly to strangers, their bark makes them good watchdogs. Some, especially males, tend to growl and snap at their owners when irritated. A good breed for novice owners and apartment dwellers. Also a good choice for considerate children.

Potential medical problems: Watch for dislocating kneecaps, corneal problems and, because of the breed's short nose, respiratory difficulties and heatstroke. Eyelid and eyelash problems often occur, as do gum and tooth problems and ear infections.

Siberian Husky

Medium-size dogs, huskies stand 20 to 23 inches at the shoulders and weigh 35 to 60 pounds. The thick coat is equipped with a dense undercoat for protection from the harsh elements of Siberia. Brushing is required twice weekly, daily when they are shedding. Even though heavy shedding occurs, Siberians are exceptionally clean dogs; they do not carry odors as other large breeds do. Most Siberians in the United States are gray or black and white. (Red and whites and solid whites also exist.) The ears are pricked, and the tail is furry and carried over the back.

Origin: These dogs are properly called Siberian Chukchi, after the Chukchi natives of northeastern Siberia. These people see their dogs as their most valuable possession: companions for their children, sled dogs and guards of their families. The first Siberian dogs were imported to Alaska in 1909 for racing purposes and are now popular as companions.

Personality: Most Siberians are good-natured, friendly and playful. Naturally very energetic working dogs, they require plenty of vigorous exercise. Siberians are bred to be outdoors in the cold weather, so that's where they are happiest. Although they will adapt to city life, some Siberians become restless and destructive with too much confinement. Natural wanderers and hunters, they must be carefully supervised when outdoors, especially if there are possible prey animals available (cats, chickens, sheep, deer). Though as a rule they are friendly to strangers, some Siberians are excessively timid. Obedience training requires patience and a gentle technique. Siberians are "babies" and will yelp at the slightest threat of pain. Housetraining is usually easy. Not the best dogs for apartment living or for novice owners, Siberians are good with children if raised with them.

Potential medical problems: Be alert for progressive retinal atrophy (PRA), hip dysplasia, cataracts, corneal disorders, thyroid deficiency and zinc-deficiency skin disease.

Yorkshire Terrier

Standing seven to nine inches at the shoulders and weighing three to seven pounds, Yorkies are toy dogs whose coats are straight and silky and need to be brushed and combed every day. Although the hair of show dogs drags on the floor, most pets are trimmed every few months. The hair on the body should be dark steel blue, with rich tan markings on the head, chest and legs. Puppies are born black. The ears prick up, and the tail is docked short.

Origin: A relatively recent breed, it first appeared as a Scottish terrier in 1861 at a dog show in Leeds, England. As the Skye terrier was also first exhibited as a Scottish terrier, it is believed that the Yorkie is a descendant of the Skye. Other breeds that are thought to have played a part are the black-and-tan Manchester terrier, the Maltese and the Dandie Dinmont terrier. Within 20 years, through selective breeding, the breed was reduced in size to a toy. It was first introduced into the United States in 1880, although the specimens varied in size from 3 to as much as 15 pounds. The breed characteristics are now more fixed.

Personality: Although it's a toy—and most of the time, a greatly pampered one—the Yorkie has not lost its natural terrier spirit. Very bright, assertive, lively, playful and clever, Yorkies do not like to be left out of family activities. They need little outdoor exercise because they usually get enough indoors. They usually like strangers but are also good watchdogs. Some are timid, and many are scrappy with other dogs. They are easy to obedience-train if gentle techniques are used. Housetraining can be difficult, and many Yorkie owners opt for permanent paper training. Many are noisy "yappers" and bark at everything. This is a good breed for novice owners and fine for apartment living. In general, Yorkies are good with considerate children, especially if raised with them.

Potential medical problems: Be mindful of dislocating kneecaps, eye infections and gum and tooth problems in these dogs.

Cats

Abyssinian

Bred to look like a small wildcat, this medium-size cat is lithe, firm and muscular with a distinctive coat. It is slender and fine-boned but not as extreme as a Siamese. The Aby has a medium broad head and softly pointed muzzle. Eyes are expressive, almond in shape and green, yellow or hazel in color. The coat is short, soft, silky and fine-textured. Daily grooming is advisable to remove dead hair, but it is easily done. Although only two colors—ruddy (normal) and sorrel (red)—are accepted in the United States for competition at this time, blues also occur naturally. All have ticked coats, with each hair containing two or three bands of alternating colors.

Origin: Although it resembles the ancient Egyptian cats seen in pictures, the Aby is a hybrid, probably developed in England (where the first Aby was recorded in 1882) from North African ticked cats. The breed came to the United States at the turn of the century.

Personality: Affectionate, very playful, quiet and good with children. Known for their high intelligence, Abyssinians can easily be trained to do tricks and to retrieve. Because of its active nature and its preference for wide-open spaces (it dislikes confinement), the Aby is not the best choice for an apartment.

Potential medical problems: Be wary of kidney disease (amyloidosis), progressive retinal atrophy (PRA), neurologic disease in young cats, skin conditions and dislocating kneecaps.

American Shorthair

American shorthairs are medium- to large-size cats with muscular bodies and short, hard coats. Little grooming is necessary. The head is large with full cheeks and a strong jaw; the legs and tail are medium in length. Over 30 solid colors and patterns, including calico and tortoiseshell, are recognized. As domestic shorthairs are basically mixed breeds, they are seen in many shapes, sizes and colors.

Origin: Experts generally agree that nonpedigreed domestic shorthairs first came to North America aboard the Mayflower in 1620. These original cats bred freely for years and kept a fairly consistent look. In 1904, the first American shorthair was recognized as a breed. Over the years, cat fanciers have worked hard to distinguish members of the American shorthair breed from mixed-breed cats. Most cats that are known today as domestic shorthairs (or medium-hairs or longhairs) are not American shorthairs. These most common family cats are primarily nonpedigreed mixed breeds.

Personality: American shorthairs are confident, intelligent and independent. They are known for their loving and gentle nature toward people. Their ancestry as survivors and hunters is evident in the personality of today's cats; that is, they adjust easily to new situations. The personality characteristics of domestic shorthairs are more variable.

Potential medical problems: Hemophilia and other blood diseases are hereditary in British shorthairs, close relatives of the American breed. Diseases of almost every organ system have been observed in domestic shorthairs, some related to coat color. For instance, tricolor cats are more likely to get mammary tumors, and white cats show some tendency toward deafness.

Burmese

Medium in size with a muscular, compact body, the Burmese has a rounded head and chest and slender legs.

The eyes are almond-shaped and colored deep or golden yellow. The coat is short, fine and glossy and requires little grooming. Until recently, the only color accepted by the Cat Fanciers Association was the original sable (brown); now blue (gray), champagne (chocolate) and platinum (lilac or pink-gray) are also recognized. Reds, creams and tortoise-shells have also been bred. The Bombay, which looks like a black Burmese, was bred from the Burmese and the black American shorthair.

Origin: Another hybrid, the Burmese was developed in the United States in 1930 from one brown, female, Oriental-type cat imported from Burma and bred to a Siamese. The resulting kittens were bred back to their mother to produce the brown Burmese cats known today.

Personality: An excellent pet; very affectionate, social, intelligent and playful. Has a Siamese voice, but tends to be less noisy. Does not like to be left alone, so it is usually better to have two Burmese.

Potential medical problems: Check for eye ailments (dry eye, cherry eye), heart defects and facial malformations.

Himalayan

A Persian body type with the same type of coat, which requires daily grooming and sheds heavily. The Himalayan body is of an even pale color, with points of a different color on the face, ears, legs and tail. This breed has the same point colors seen in the Siamese. The most common are seal point (seal brown points, fawn to cream body); chocolate point (milk chocolate–colored points, ivory body); blue point (slate blue-gray points, light blue-gray body); lilac point (pink-gray points, white body); red, or flame, point (orange to red points, creamy white body); tabby, or lynx, point (striped points of any color with appropriate body color). Eyes are round and blue.

Origin: A triumph of hybridization, the Himalayan, or colorpoint, is essentially a Persian with Siamese coloring. It is not a Siamese with long hair, nor does it come from the Himalayas. The name comes from the Himalayan rabbit, which has a similar coat pattern. It took ten years of intensive breeding experiments to produce the Himalayan.

Personality: Affectionate, devoted and intelligent. Good with children. More active and playful than Persians.

Potential medical problems: Check for cataracts, skin conditions and kidney disease; also susceptible to diseases common in Persians.

Maine Coon Cat

A medium to large cat with a muscular and broad-chested body. The head is small and set on a powerful neck. The coat is heavy and shaggy, shorter on the face and shoulders and longer along the back, stomach and hind legs. The beautiful tail of the coon cat is furnished with long, flowing fur. Unlike other long-haired breeds, the coon cat requires little grooming. The brown tabby is traditional, but Maine coons have been bred in a variety of colors and patterns, including solids, parti-colors and tortoiseshells. Eyes may be green, gold, copper or blue.

Origin: Considered by many to be a "natural" breed that originated in the northeastern United States and Canada. Most likely, it developed from accidental matings between domestic shorthairs and longhairs that were brought by traders from Asia Minor to Maine and other parts of New England. The name *coon cat* comes from the breed's resemblance to a raccoon. It is the first pedigreed American cat breed and won "best cat" at the first large American cat show, held in 1895.

Personality: A hardy and active breed, the coon cat is known for being a good mouser. It is playful and good with children but may be shy. It is quiet and has a soft, chirping voice. Although able to adapt to both indoor and outdoor life, the coon cat prefers plenty of space to roam.

Potential medical problems: Look out for flattened chests in newborn kittens, dislocating kneecaps and hip dysplasia.

Manx

Generally considered tailless, the Manx is actually seen in three varieties: "Rumpy" has no tail at all; "stumpy" has a tail stump one to five inches long; and "longie" has a complete tail. (Only the tailless Manx is accepted for shows.) The completely tailless variety is similar to the common domestic shorthair but has a hollow where the tail would be. Medium in size, with a solid body, rounded rump, muscular hindquarters and a short back, which arches from shoulder to rump. The head is large and round, and the eyes are large, round, expressive and of any color. The coat is short, with a thick and cottony undercoat. Daily grooming is recommended. All colors and coat patterns occur.

Origin: The Isle of Man in the Irish Sea is recognized as the home of the Manx. Tailless cats have been known to exist for centuries in Malaysia, Russia and China, so it is possible that some were brought to the Isle of Man by ships. Once the animals were there and isolated, the gene for taillessness spread throughout the island's cat population.

Personality: The Manx cat is intelligent, courageous and affectionate. It is a good hunter and mouser and is good with children and dogs.

Potential medical problems: A unique concern: If the Manx tailless gene is inherited from both parents, it is lethal in the womb. These cats are also susceptible to spina bifida (or malformation of the vertebrae and spinal cord), which causes ailments of the large intestine (including incontinence and constipation) and hind limb troubles. Also watch for other spinal cord defects, rectal problems, corneal problems and skin infections.

Persian

Medium in size with a cobby, or compact, body type. Persians have relatively short legs and are deep-chested and broad across the shoulders and rump. The face is flat and round with a snub nose. With its long, thick coat with fine-textured hair, the Persian requires daily grooming. Shedding occurs all year round. The Cat Fanciers Association registers 51 separate colors and color combinations of Persian cats. Himalayans are classified as Persians by the association, but other cat registries consider them a separate breed.

Origin: One of the oldest breeds, it is considered a "natural" breed by many, as opposed to those breeds that are "man-made," or hybrid. Although their exact origin is unproven, some believe that the modern Persians are descended from Turkish Angora cats, which were crossed with other long-hairs from Iran (Persia), Afghanistan, Burma, China, Russia, France, Italy and England. Selective breeding over the past 100 years has produced the modern Persian type in the numerous color varieties seen today.

Personality: Affectionate, undemanding and good with children, Persians tend to be less active than other types of cats. Their heavy bodies and short legs also make them less athletic. Though long-haired cats are more prone to litter box problems than shorthairs, generally speaking, Persians are a good choice for apartment living.

Potential medical problems: Watch out for hair balls, breathing problems, excessive tearing, eyelid abnormalities, glaucoma, progressive retinal atrophy (PRA), skin conditions, urinary stones, kidney disease, dislocating kneecaps and hip dysplasia.

Siamese

This medium-size breed is svelte and long, with long legs and fine bones, but strong and muscular. The "proper" head shape is also long and narrow, although Siamese with round heads are not uncommon. Eyes are almond-shaped and blue. The coat is short and fine and requires little grooming; it sheds, but not excessively. Siamese have pointed coats with pale bodies and darker colors on the face, ears, tail and legs. The Cat Fanciers Association accepts only four colors: seal point (seal brown points, fawn-colored body); blue point (deep silver or slate blue points, gray body); lilac point (pink-gray points, white body); and chocolate point (milk chocolate points, pale ivory body). The Siamese cats with points of other colors (red, tortoiseshell, tabby) are classified as colorpoint shorthairs. Crossbreeding has resulted in many popular breeds, such as the Balinese, Havana brown, Himalayan, Oriental shorthair and Tonkinese.

Origin: Although its exact early history is unknown, many consider the Siamese a "natural" breed. Its origin is certainly Asian, possibly from Siam (Thailand) or Burma. According to legend, ancestors of the breed were kept by the kings and priests of Siam and were trained to guard the royal palaces and temples. It is believed that the first Siamese cats ever exported were a gift to the British consul from the king of Siam in 1885. The Siamese is the most popular pedigreed cat breed in America, second only to the unpedigreed domestic shorthair.

Personality: Good companion—intelligent, very active, very demanding and very vocal. Siamese may be intolerant of restraint and inconsiderate children. Although Siamese are among the most playful and social of all breeds (many enjoy walking with their owners on harness and leash), some people find them too noisy.

Potential medical problems: Beware of hair loss, kinked tail, crossed eyes or rapid lateral movement of the eyes (neither impairs vision), neurologic disease, intestinal problems (including cancer), heart defects, breast cancer, asthma, skin diseases and tumors, glaucoma, tooth problems and hip dysplasia.

Index

Note: Underscored page references indicate boxed text.